FUBAR 2.0

A Soldier's Insight into Military Chaos

Gary W. Clark

FUBAR 2.0: A Soldier's Insight into Military Chaos

www.Fubar2.com

Published by PhotoTree.com

9 8 7 6 5 4 3 2 1

ISBN: 978-0-9835785-7-4

Cover and interior design by Gary W. Clark

Edited by Gena Philibert-Ortega

Dedication

This book is dedicated to Jay Rodney Goodrich, my closest friend while growing up in a working class neighborhood in Wichita, Kansas. Jay was seriously wounded in Vietnam, a sucking chest wound if I remember correctly. He survived and lived a good life full of great friends, though he died too young of a nerve disease. Being another non-nonsense guy, he would appreciate all the FUBAR stories I wrote about.

Special thoughts go out to Mike McDonald and J, D. Arnett who died within a month of arriving in Vietnam, both on their first mission as we were told. I went out on the town with both of them and friends the night before each left. Also, to Johnny Guillen, who I played a lot of basketball with on the asphalt court of St. Patrick's church parking lot.

In addition, I offer respect to all my friends and others I knew from the neighborhood and school who died and served. There were a lot of them.

James (J.D.) Arnett ...... KIA, 9/15/68, 9th Infantry
Michael McDonald....... KIA, 12/21/68, 9th Infantry
John Guillen .... KIA, 2/2/70, 11th Infantry
Carl (Leo) Ornelas .... KIA 2/5/68, III Marine Amphibious Force
Jay Goodrich .... U.S. Army, Died 3/11/2011
Darrel Beth .... U.S. Army, Died 12/15/2010
Harold Coleman......... U.S. Army
Leonard Foster .... U.S. Army
Mike Maddox .... U.S. Army
Jerry Reynolds .... U.S. Marines
Jerry Schroeder .... U.S. Marines
Ronald McEwen .... U.S. Army

Prologue

Chaos is not planned. It is usually a by-product of scale or haste. In times of conflict, neither increased scale nor haste are easily avoided. The military may need to add thousands of new troops, and activate or create new units. These missions require completion ASAP.

FUBAR is a by-product of chaos. FUBAR recognition is not always bad; sometimes it shines a light on a problem, giving innovative young men and women the opportunity to develop a solution.

A classic case of the average soldier solving a FUBAR situation occurred during World War II. The problem was impenetrable French hedgerows that were delaying allied tank advancement after D-Day. These decades old root-encrusted hedgerows were solid dirt berms as tall as a man and several feet deep; they could stop a tank from driving through or over them. The solution was the invention of tank teeth. A couple of farm boys came up with the idea of welding jagged steel teeth to the front of tanks. The teeth could penetrate the hedgerow and the tank would plow through the obstructions. Being farm boys, they knew about plowing tough soil. Their solution immediately became tank modifications throughout many armor divisions.[1]

Not all FUBAR challenges are avoidable or solved. However, many do have an immediate and sometimes lasting impact on service men and women. Relentless FUBAR events may teach someone how to be innovative. They may teach how to look beyond an apparent problem and find the root cause. All FUBARs are teaching moments.

Many veterans do not tell their stories; this is especially true of the 'greatest generation.' However, this author suggests their experiences, especially FUBAR moments, shaped them in countless ways, many for the better.

See if you recognize the affect of FUBAR moments on yourself, a loved one, or friend.

Table of Contents

Introduction

Captain Miller:

We all have orders, and we have to follow 'em. That supersedes everything, including your mothers.

Corporal Upham:

Yes, sir. Thank you sir.

Private Reiben:

Even if you think the mission's FUBAR, sir?

Captain Miller:

Especially if you think the mission's FUBAR.

Saving Private Ryan (1998)

In the above banter, Captain Miller, played by actor Tom Hanks, offered an eloquent response to a conversation challenging the mission's value.

The Army Rangers' mission in this movie was a clear FUBAR example to most of the squad, risking the lives of eight men to save the life of one. This pronounced reference to FUBAR in the conversation probably resonated with a vast number of soldiers in the audience, both young and old.

The millions of other people who watched the movie probably missed the meaning of that sequence. The problem is not the cryptic meaning of the word FUBAR, it is the deeper consequence of FUBAR that most people miss. If soldiers tag an event as FUBAR, this usually means there was extreme frustration, aggravation, or just plain futility with the situation.

This book is about the many FUBAR moments during my two years of draft-inducted enlistment during the Vietnam War. My experience was those FUBAR moments seemed to occur once a day. That is probably an exaggeration, and to make it clear, none were of the magnitude and dangerous nature experienced by many servicemen and women.

~~~

This year, 2014, marks the 50th anniversary of the Vietnam War's official beginning, or as close to official as can be historically chronicled. The year 1964 gave us the Gulf of Tonkin Resolution, passed by Congress and enacted on August 10, 1964. This resolution was in response to the storied attack by North Vietnam torpedo boats on the destroyer USS Maddox.

The resolution was only 355 words long, although it set the stage for one of the most controversial actions by the United States in history. The simple, if not benign and certainly vague resolution title could not begin to predict the forthcoming impact on the United States. The title read:

> *To promote the maintenance of international peace and security in Southeast Asia.*

After three short paragraphs describing the attack on the USS Maddox and the United States' responsibility and goals, the resolution gave unprecedented power to the President without actually declaring war.

> *Resolved by the Senate and House of Representatives of the United States of America in Congress assembled, That the Congress approves and supports the determination of the President, as Commander in Chief, to take all necessary measures to repel any armed attack against the forces of the United States and to prevent further aggression.*

While this is not a history of the Vietnam War, all the stories in this work would not have happened if the War had not progressed as it did. I brought forward some of the history for context.
~~~

Why FUBAR 2.0? World War II and the Korean War also gave us FUBAR; in retrospect, they gave us FUBAR 1.0. The Vietnam era produced 2.0. We recently have experienced 3.0.

The purpose of this book is to be entertaining and funny, much at my expense; unveil some shocking military faux pas; and open a window to the experiences of veterans you know, past and present. This may help illuminate the situations they encountered, many of which shaped their character.

The book does not disparage the military, soldiers, sailors, Marines, or Air Force personnel. It is not anti-anything: war, military, or government. Neither is it hawkish, philosophical, or political. Still, this book does not shy away from history and truth; many government, political, and military decisions and actions are referenced to illustrate the FUBAR concept. Common sense, good judgment, fair play, and diplomacy will always rule.

~ ~ ~

My tour of duty in the U.S. Army from February 20, 1969 to February 13, 1971 placed me in Ft. Polk, Louisiana; Ft. Sill, Oklahoma; southern Germany; and South Vietnam. I experienced the proverbial "growing up" during this time as I entered service at the naïve age of 19 years.

I did not see combat in Vietnam; the stress, hardships, and tragedies experienced by those who did are forever in the minds of all of us. Many fine books have been written about the battles, firefights, and life in the boonies. FUBAR 2.0 looks at the overriding nature of the military and its affect on everyone that served, not matter what position and rank was held, or on which posts they served.

~ ~ ~

On a final note, the stories here are all true; I have tried not to embellish them – too much. I always thought it was amazing all these tales happened to me or I observed them. Now, I know I am not that unique, so this tells me there are thousands, no millions of FUBAR stories in the minds of others, and clearly, many are more up-to-date than mine, as a new generation of young men and women have experienced the rigors of the military and war over the last twenty years.

I encourage others to write about their own versions of FUBAR. Maybe one of you will publish FUBAR 3.0.

Gary W. Clark, Spec 4, US 54566721
101st Airborne Division
Camp Campbell, Phu Bai, South Vietnam 1970 – 1971

Chapter 1 – FUBAR History

FUBAR – SNAFU – TARFU: a common theme runs through them.

~ ~ ~

For the uninitiated, I will blurt out only one time the definition of FUBAR, or at least what the letters stand for: **F**ucked **U**p **B**eyond **A**ll **R**ecognition. Ok, I said it. I will not use F*** to gently refer to the politically incorrect word, nor will I blatantly throw F-bomb grenades. FUBAR by itself is self-explanatory, even if it is something to be snickered at.

~ ~ ~

The term's origin is attributed to a number of sources. I like to think it was just good ole GI ingenuity. (See, another acronym.) GI means 'Government Issue.' Everything a soldier possessed was issued to him by the government, so he became a GI himself.

Military units and now the high-tech industry, love acronyms. If the military invented them, the soldiers perfected them. FUBAR became a common if not overused slang during World War II. The R sometimes referred to Repair or Reason.[2] Our catchy term probably has no traceable origin. A 1944 Army animated short film titled The Three Brothers included characters named Fubar, Snafu, and Tarfu. Both SNAFU and TARFU are simply variations of the same theme: **S**ituation **N**ormal **A**ll **F**ucked **U**p and **T**hings **A**re **R**eally **F**ucked **U**p.

This silly movie attempted to convince a soldier that even his most boring job was important to the war effort. The original animation is available for viewing on YouTube. No doubt, by the time they filmed the 1944 movie, the unflattering names were already well entrenched in soldiers' vocabularies. While SNAFU has endured the test of time and is occasionally heard, TARFU has gone the way of the horse cavalry.

FUBAR has been used in serious productions also. A seven-part documentary by Ken Burns on World War II aired in 2007, with the fifth episode titled FUBAR: September 1944 – December 1944. This part covered two campaigns, the disastrous Market Garden operation in Holland, and the high-casualty battle for Peleliu Island by U.S. Marines in the Pacific theater.

Beyond the name of the episode, frank language throughout the series caused concern for many PBS stations airing the documentary. Hoping to avoid FCC meddling over language and prevent complaints by viewers, PBS supplied both a *clean* version of the series, and one with original dialogue to their 350 affiliates. The local stations could decide which version to air. [3]

Sadly, as a documentary, the clean version would not show real life that occurs under the umbrella of war.

~ ~ ~

Acronyms were both official and unofficial. Jeep came from the phonetic sound of GP or General Purpose. General Purpose Vehicle was the official name but it was much easier to say 'jeep.' HU-1 the early designation for the Vietnam mainstay helicopter from Bell Helicopter became the 'Huey.' Later the chopper designation became UH-1, with the 'U' meaning utility and of course the 'H' representing helicopter.[4] The Army-assigned name 'Iroquois' never stuck with ground pounding soldiers who were dropped into hot LZs (landing zones) by Hueys.

If the military comes up with a long name or description for something, the guy in the dirt will probably give it an acronym or bastard name. The mess hall (dining area) was quite frequently a source and target of many nicknames. 'Chipped Beef on Toast', a breakfast staple, became SOS or 'shit on a shingle.' The first sergeant, the top enlisted man in a company, becomes 'Top.' A Marine gunnery sergeant becomes 'Gunny.' Give a soldier, sailor, marine an official name, description, or acronym and he will change it or make one up that better fits the situation.

FUBAR is an ingenious term that has survived the test of time.

Besides enjoying the fifth grader snicker factor of saying the word, FUBAR lets you cuss without getting into trouble. It provides a great way to use a F-bomb without incurring the wrath and furrowed-brow looks of kinder and gentler people. Hollywood has figured this out, the term is used frequently in movies and television; it helps keep the pesky 'R' rating away from a movie aimed at teens.

More important than shock value, is that FUBAR succinctly describes a systematic theme that runs through many military endeavors. To illustrate what this means, you only need to look back on the television series M*A*S*H, and movies such as Catch-22. Both of these entertainment hits created terrific story lines from preposterous situations worthy of the FUBAR stamp. The eleven year run of M*A*S*H proved that nearly an infinite number of FUBAR events were available to write hundreds of scripts about. Yes, M*A*S*H could have been titled F*U*B*A*R.

The movie *Catch-22* gave us the best FUBAR logic ever. This epic World War II comedy about U.S. bomber pilots stationed in England tried to explain that it was crazy to fly raids over Germany, and therefore if you were crazy, you should not have to fly them. The script went:

> *"There was only one catch and that was Catch-22, which specified that a concern for one's own safety in the face of dangers that were real and immediate was the process of a rational mind. Orr [Captain Orr] was crazy and could be grounded. All he had to do was ask; and as soon as he did, he would no longer be crazy and would have to fly more missions. Orr would be crazy to fly more missions and sane if he did not, but if he was sane he had to fly them. If he flew them, he was crazy and did not have to; but if he did not want to, he was sane and had to."*

> *"Yossarian [Captain Yossarian] was moved very deeply by the absolute simplicity of this clause of Catch-22 and let out a respectful whistle."*

You cannot argue with FUBAR. It just exists.

Chapter 2 – The Draft and Induction

"You have great test scores and are highly qualified for a helicopter avionics position. Just sign here for an extra four years."

~ ~ ~

As an 18 year-old in 1968, I was not attending college, had a dead-end warehouse job, and I was in perfect health. My near term future was easy to recognize.

By the end of that year, I had the sorrow of having seen two school friends killed in action, two close neighborhood friends seriously wounded, and several friends who already returned from their time in Vietnam.

The increasingly unpopular Vietnam War was in full swing and my draft classification was 1A – it does not get any higher or more draft-eligible than that.

~ ~ ~

Richard Milhouse Nixon was running against President Lyndon Baines Johnson for President of the United States in the fall of 1968. LBJ was running for re-election, it would be his second full term after taking office upon President Kennedy's assignation in November of 1963, and then winning the 1964 election.

President Johnson's administration had vigorously pursued the Vietnam War, escalating it year after year. History shows that 1968 recorded the War's highest yearly troop count and U.S. death toll.

Nixon was using the quagmire of the war and a demonstration-plagued opposition from the home front as prime campaign rhetoric. His 1968 Presidential nomination speech took to task the in-office democrat president:

> *For four years this Administration has had at its disposal the greatest military and economic advantage that one nation has ever had over another in any war in history.*
>
> *For four years, America's fighting men have set a record for courage and sacrifice unsurpassed in our history.*
>
> *For four years, this Administration has had the support of the Loyal Opposition for the objective of seeking an honorable end to the struggle.*
>
> *Never has so much military and economic and diplomatic power been used so ineffectively.*
>
> *And if after all of this time and all of this sacrifice and all of this support there is still no end in sight, then I say the time has come for the American people to turn to new leadership -- not tied to the mistakes and the policies of the past. That is what we offer to America.*
>
> *And I pledge to you tonight that the first priority foreign policy objective of our next Administration will be to bring an honorable end to the war in Vietnam. We shall not stop there -- we need a policy to prevent more Vietnams.*

To paraphrase Richard Nixon, he basically said "if Johnson couldn't get the U.S. out of the Southeast Asia war in six years, he didn't deserve to be president."

While I was not eligible to vote yet, I still felt Nixon was my kind of guy and rooted for him to win the presidency – which he did. I was in Vietnam 22 months later, assigned to the 101st Airborne Division. FUBAR.

~~~

It would have been too optimistic to think that if Nixon did win the election in November of 1968 that the war would end overnight. Nevertheless,
~~~

hope was a quicker path to withdrawal would be in the works. Richard M. Nixon was sworn in as President of the United States on January 20, 1969, Thirty days later, February 20th, I was drafted. FUBAR.

~ ~ ~

I reported to the Pine Bluff, Arkansas federal building on February 20, where they then put me and my fellow band of scruffy draftees onto a bus bound for the Little Rock Induction Center some 50 miles away. Since Pine Bluff was my home when I registered for the draft, I was selected by that local draft board, and had to report to their induction center.

In retrospect, I identify with Bill Murray's and Harold Ramis' characters in the movie *Stripes*, as they made their way through an alien maze of induction processes and bored workers. Get on the bus, get off the bus, always have your manila envelope with you; lose that and you are a leper that no one knows what to do with.

If nothing else, the military is quite efficient at some things. An enormous administrative machine processed 45,000 draftees across the country in February of 1969. That was one of the highest draft quotas of the war.

Induction center personnel perform health exams, monitor fitness and dexterity tests, and give draftees written exams to determine their fitness and capabilities. Just because you were drafted did not mean you were in the service yet. At this point in the process, a draftee could be rejected for any number of reasons. However, short of heart murmurs, flat feet, or a missing trigger finger you probably made the grade.

When the battery of tests were complete and graded, a short interview was conducted with each draftee. Short is being kind, they were not interested in your opinion of your strengths or weaknesses. They would determine those. This was more like a speed dating session with someone who has heard all the lines and did not really want to be there. Actually, the interview was sort of an enlistment pitch because they would look at your test scores and education level to see what job in the Army you were best qualified for. I scored pretty well on all tests and our conversation progressed to military occupations. He offered a couple suggestions and we agreed that helicopter technician was an appropriate MOS (Military Occupational Skill) for me. Well, I was now enthused. He would put me down for helicopter repair school as my preference after basic training.

Then the catch, nothing is certain or free with the Army. Even if it is promised, they have their fingers crossed behind their back. Yes, I was qualified to attend helicopter repair school, and yes, he wrote that down as my preference. Remember, I was drafted and nearly all draftees go straight to infantry training or to other combat arms like artillery or armor after basic training. Even a few draftees were sent to the Marines when they did not meet there volunteer quota.

Draftees were only called to active duty for two years. To ensure I would go to helicopter repair school, I would need to give them four additional years for that training. That was probably fair, but at 19 years of age, I was not ready to commit any more of my precious youth to the Army than was necessary.

I did not enlist for the extended time and instead drew infantry. FUBAR.

~~~
~~~

Chapter 3 – Basic Training FUBAR

Formation is an opportunity for the first sergeant to tell you how badly you screwed up that day.

~ ~ ~

There is no a better place to be engulfed in the FUBAR world than basic training. Some of it is petty in the larger scheme of things: like having your head shaved, yelled at nearly every moment of your waking day, and no free time.

As training progresses, you accept those inconveniences and move on to coping with specific FUBARs, some tailored just for you, or so it seemed.

The military is based on the management technique of everybody doing the same thing, the same way, with the same results. Variations are not tolerated. That concept is easily recognized by shaved heads, standardized uniforms, and everyone walking, standing, sitting, and talking the same way.

This is where your personal FUBAR moments can sneak up on you and wreak havoc with your confidence, body, and even pay. If you stray from standard protocol, you will stand out and then may the Lord have mercy on your miserable trainee life.

Now I was always a bright kid, though my report cards frequently noted I 'didn't perform up to my potential.' Never did think I was a screw-up; hell, I was even an Eagle Scout. However, I did step in it numerous times during basic training.

~ ~ ~

We had a crusty first sergeant, actually I think that is redundant, who struck terror into all trainees, especially us skinny ones. I was only 5 foot 9 inches tall, and weighed a whopping 130 pounds when drafted. This was not exactly the ideal prototype for the next generation killing machine. Hey, it was the Army's idea for me to be there.

My first dumb luck included being assigned to the first squad of the first platoon. The means while standing in formation, which happens several times a day, I am on the front row in the first platoon. Now in a training company there were 200 trainees in five platoons with four rows of squads each. The platoons are next to each other.

Formation is the military term for gathering everyone in a company in one area so the first sergeant to tell you in the morning what you are going to do that day, then at the end of the day he tells you how badly you screwed up the day's training.

It was only March, so the heat did not get to me, but after a particularly fast and long march, I was feeling queasy in formation. We are standing at attention in formation getting our end-of-day berating and here it came – my lunch that is. I tried to keep it in, but it was determined to make an appearance. In my best military bearing, still at attention, I stiffly leaned forward and threw up right in front of God, 200 men, and the first sergeant. FUBAR.

His actions were civil, not harmful; he simply ordered one of his squad leaders to "get that man to the back of the formation." I think he was a surprised as I was. Ok, all was cool, after formation I only had to use a shovel and clean up my untimely deposit. There were no direct repercussions.

However, to this day, I firmly believe I made a lasting impression on the grizzled first sergeant. It's a very good bet it was not a favorable impression however. The best tactic in successfully navigating basic training is to be invisible. FUBAR.

~ ~ ~

Before ATMs, debit cards, even direct deposit, the Army paid in cold hard cash once a month. It was a pretty formal and secure event. The company pay officer along with a ranking sergeant would go to the post headquarter and pick up the pay vouchers, one for each soldier and a bag or bags of money. Each of them armed with .45 caliber automatic pistols on their belts. A table was set up in company headquarters with several officers and sergeants to process each soldier as they came from a long line snaking around the company area. One would check your ID, the next would have you sign the pay voucher, and the next two would count out your money.

Now the service can be very confusing to those of us who over-think things. For example, when any kind of roll call is conducted, you always reply with your last name first, then your first and then middle initial.

So I got used to my name being Clark, Gary W. When it came time to sign my voucher, I signed 'Clark, Gary W.' Guess what? That is not a signature! The signature should be 'Gary W. Clark.' What I signed did not match what the voucher said. And in the Army you cannot color outside the lines.

Guess who was supervising pay distribution? The crusty first sergeant. He exploded with disbelief; yelling that was the dumbest thing he had ever seen (which I doubt) and they could not pay me. What is worse, it made extra work for him, which you never want to do; he had to request a new pay voucher from headquarters, the company cash did not balance now, and it would take a couple days to straighten out the paper work.

If the grudge-bearing first sergeant did not have my number from throwing up in formation, he certainly did now. FUBAR.

~ ~ ~

So, I think that little payday escapade eventually cost me money. Like I bragged about before, I was a bright kid, in great shape for a skinny kid, knew all about first aid from Boy Scouts, and gained experience with a rifle while hunting with my dad.

I earned some of highest scores on written test. Excelled on PT (physical training) tests, and qualified expert on all weapons training. Being a skinny kid actually helped me in physical tests: I could do more pushups and pull-ups than most, and run through obstacle courses quickly. I played basketball relentlessly as a kid, and was even a pole-vaulter in school; so I did have good physical dexterity and skill. Of course, I still looked like a dweeb.

Now why is all of this important? Even the trainee who scored last on all the tests was still going on to his next training school.

Here is where the money comes in. As a draftee or volunteer, enlisted servicemembers enter the service with an E-1 pay grade. This is bottom of the ladder in pay and respect. As you are promoted, the pay grade progresses to E-2, E-3, and so on. Officer pay grades are designated with an O, such as O-1 (second lieutenant) and on. Basic training was eight weeks, and then you usually went to another eight weeks or more to your MOS school. Remember, MOS is the Military Occupational Skill. Most advanced training schools were eight weeks unless they were highly technical – like helicopter repair school – which I would never see.

It was most likely you would not receive your first promotion, to E-2, until graduation from the MOS or advanced training school. Except, if you were one of the chosen few, chosen by the first sergeant. You see the problem coming up?

A small percentage of basic training graduates were promoted to E-2 at the end of those first eight weeks. It was used as incentive for trainees to try harder. I was working on that. Moreover, the sooner you made E-2, then the sooner you were eligible for E-3, and further ranks. The affect was like compound interest.

From the beginning, we were told promotion selection was based on how well you did on exams, skill tests, physical training scores, and marksmanship. Hell, I was near the top of the list on all of them. Except for the first sergeant factor. I was too visible – negatively visible. Truth is, no matter if the sun does come up in the east every morning, if the first sergeant say it does not then it does not. Ignoring facts like test and performance scores was no problem. If the first sergeant would have been asked to describe Clark, I am sure part of his answer would include the phrase 'screw-up.'

Guess who did not get a promotion out of basic training? FUBAR?

~~~

A military branch itself is not directly responsible for many FUBAR moments, but it creates surrounding environments that affect a soldier's life, both in performance of their duty and the private life they try to establish or maintain such as starting and raising family.

It was nearly a certainty that if you were drafted in the mid and late-1960s, you would become an infantryman, followed by deployment to Vietnam. This inevitability was the cause of an often seen affect.

One of the phenomena that militaries have recognized is that an overwhelming number of service members get married right after basic training or just before deployment into harm's way. The services actively discourage these marriages; statistically the hasty unions experience a high rate of failure.

We were given numerous classes and speeches from the company commander, chaplains, and even marriage counselors on why not to marry at this time. I do not remember their exact reasons and talking points, though I was one that should have listened.

While I am not a psychologist, nor play one on TV, I have come up with several reasons why this trend continues. First, the military takes everything away from a recruit. Besides your hair, clothes, and your car because you cannot afford it, you have nothing to call your own. If you have a wife (or husband) you now have a piece of life that the service cannot take away. You are now an independent person again, not really, but it does add back to your unique identity.

The other reason may seem darker, but I can guarantee it is a reason for
~~~

some military marriages. During times of war, many service members know they are going into dangerous situations where they may not come out. This is especially true of the large number of soldiers in the combat arms branches of the Army and Marines. More than one soldier has admitted, if only to themselves, that they did not want to die without ever experiencing marriage. I have no statistics, but living within a couple miles of the Camp Pendleton Marine base in Oceanside, California for the last 12 years, I have observed that the trend of early marriages continues.

This is not to say that all early military marriages were disingenuous or missing a two-way connected love. Congratulations to all of those that have succeeded.

This soldier knew he was going to Vietnam as an infantryman. I married a high school sweetheart within days of graduating from Advanced Infantry Training at Fort Polk, Louisiana. The marriage ended within a year after discharge from the Army. FUBAR.

Chapter 4 – Advanced Individual FUBAR

"Croak, croak, croak, I'm a tree frog, sorry I lost my weapon."

~ ~ ~

AIT, an acronym for Advanced Individual Training, was an eight-week course on infantry tactics and skills. At Fort Polk, the Army was graduating 1,000 infantrymen a week, with the high percentage of them going straight to Vietnam after a 30-day leave. Ft. Polk was one of several Infantry schools across the country, each doing the same thing.

This aged army post was a World War II era training center, chosen probably for its inhospitable terrain and climate, and the open space available for training. At one point in 1943, over one million soldiers in 19 divisions were training at the fort.

During the Vietnam era, the camp included two distinct training centers, South Fort and North Fort, about five miles apart, separated by forested hills. The South Fort was modern by Army standards; however, North Fort still consisted mostly of dilapidated World War II era barracks. After the Korean War, all of Fort Polk was placed into then pulled out of mothball status several times; and once again to train Vietnam-bound troops. It continued to service its primary purpose well, training more infantrymen for service in Vietnam than any other base in the United States.

Sand hills, swamps, rainy weather, and humid summers are some of the area's finer attributes. This was a virtual free-range reptile zoo area with poisonous snakes such as pygmy rattles snakes, copperheads, and coral snakes ever present; even an occasional alligator was found around some lakes. There is a FUBAR story about that soon.

~ ~ ~

Advanced army school assignments were announced for each trainee at the conclusion of basic training. It was an important day; your life might depend on the assignment. A drill sergeant posted the lists on a bulletin board while two hundred men crowded around to discover what their army career would be, for at least the next couple of years.

Not all advanced school were at Fort Polk; if you drew clerical training, mechanics, artillery, armor and others, they would toss you on a plane and away you went. No leave, no going home. Next stop was another Army post.

For the unlucky ninety percent, our trip was to the other side of Fort Polk for infantry training. It was not even a bus ride. The common mode of transportation for large groups of trainees was a cattle car – literally. It was as semi tractor-trailer rig with sidewalls on the trailer, no roof, benches built along the inside walls, and a row of seats down the center. For our riding pleasure, I think the truck drivers loved to see how fast they could take the curves, throwing us on top of each other from one side then to the other. The Army training process provided very little comfort or dignity at this point.

Arriving at AIT, I was still angry about two subjects from basic training. One was the non-promotion to E-2. Besides the lack of extra money, an E-1 is still a trainee, not even worthy of the rank of Private. The second irritation, which is no surprise, was the fact I did not get helicopter technician for my school. The chances were slim to none, but I held out hope, which was quickly squashed like a Louisiana mosquito.

Advanced infantry training was more serious and harder than basic. After all, next stop was the rice paddies, jungle, or mountains of Vietnam. Therefore, the drill instructors took their job serious. However, many of the training methods were based on intimidation, threats, and punishment. Punishment in front of others was the preferred method of making a point. To me, this seemed a bit leftover from the inquisition of the Middle Ages.

~~~
~~~

While I still had many embarrassing opportunities in front me, I did manage to avoid a particular indignation the drill sergeants loved to lay on trainees.

We spent most of the days training in various wilderness areas on the fort. It encompassed nearly 200,000 acres, which provided a wide expanse of varied terrain. Still, when transported, or marched, to a training area, that site had bleachers for lecturing, shooting positions built into the ground, maybe an obstacle course, and a dining area. The dining area was crude; 2" x 12" planks mounted flat on top of four-foot poles where you ate from a mess kit. Lunch was brought in by truck and served by gnarly cooks from hot and cold containers. Just like us, they didn't want to be there, especially if it was raining, which during Louisiana springtime was quite frequent.

For drinking water during the day, we filled our canteens from a lister bag. The lister bag is an ingenious contraption, though very simple. A product of World War I or even before, it is a heavy canvas bag, about 35 gallons, hung from a three-pole stand about two feet off the ground. Training area personnel hung empty bags in the morning, and water trucks would drive around the areas filling them.

Ringing the bottom, several water spigots were built into the bag, maybe as many as 10 or 12 encircling the gravity-driven drinking fountain. Soldiers could fill their canteens, usually missing the mouth of the canteen and splashing water onto the ground. The spigots would always drip, and the lister bag would sweat. These three things conspired to create a puddle of mud directly beneath the bag. Evil drill sergeants loved these types of opportunities to devise new means of punishment.

With over two hundred soldiers on a training course, law of averages says someone is going to screw up nearly every day. Infractions could be as minor as talking in formation, sneaking a smoke, falling asleep, cheating at some task, or any other number of offenses. The mud puddle beneath the bag gave the drill sergeants a venue to inflict pain, misery, and embarrassment onto their victim. The 'dying cockroach' is born.

Take one guilty soldier, make him lay on his back underneath the lister bag, and have him hold his arms and legs up as high as he can; only his back and butt on the ground. How humiliating. And, with the threat of further punishment, maybe three days of KP (kitchen police), he is warned to not lower his limbs. The arms and legs start aching shortly; the victim is not sure which is worse, the misery of situation or the pain. To help aggravate the embarrassment, the drill sergeant would have the guilty recruit chant something about how sorry he was for his wrongdoing or, maybe have him repeatedly cry out "I'm a dying cockroach." I do not know what sound dying cockroaches make, though I was about to become an expert on tree frog sounds. FUBAR.

~~~

Back to my own personal hell-story, it is just as bad as the dying cockroach, although without the physical pain.

Many training exercises involved simulated assaults on enemy positions in hilly terrain. We charged an imaginary enemy with M-16 rifles, screaming banshee-like cries of death. The enemy, drill sergeants in hidden positions, would throw flash-bang grenades and fire blanks from a variety of weapons. Instructors would wander through this simulated battle, grading the trainees on their actions.

One particular weapon the enemy instructors used was an actual live munition: tear gas canisters or grenades. They would occasionally throw real CS gas grenades at the trainees to test their ability to don their gas mask while under attack. CS is the acronym for the gas that is commonly referred to as tear gas. Though its chemical name is 2-chlorobenzalmalononitrile, the letters CS come from the gas inventors last names: Ben Carson and Roger Staughton. In basic training we trained for exposure to CS gas; the instructors crammed a group of men wearing gas masks into a building and released the offending gas. After standing in the chamber a few minutes, we took the masks off and stayed there for 30 seconds or so, trying desperately to hold our breath. If holding your breath failed, the pain was excruciating.
~~~

However, that was in a static, simple scenario with the mask already in place before exposure. In the field training exercise, our gas mask was in a pouch slung over a shoulder. Should someone started yelling 'gas', you immediately stopped, took cover, got the mask out of the pouch and put it on. Simple, yes? No!

I mentioned earlier that we were in hilly terrain, with plenty of places to hide from simulated enemy fire. Well, this was not the instructors' first rodeo; they knew where the trainees would cluster.

I was in a narrow ravine, making my way to the 'enemy' position. Much to my horror a gas grenade landed right in front of me and went off. They do not explode with flying shrapnel, only thick clouds of gas start emanating from the canisters. There was no time to hold my breath or put the mask on; I immediately inhaled full-strength gas into my lungs. Not some scattered gas that had wafted across the hills and lost potency, but concentrated terror straight out of the canister. It has been quite a few years, but I remember the pain well. The gas locks up your lungs; no breathing in or out. It burns your eyes, mouth, nose and exposed skin. If you have crotch rot from lengthy marches, it really burns there. It certainly does what it is supposed to do – incapacitate an enemy soldier.

I struggled to get out of the ravine and scramble away from the direction the gas was blowing, I never did get the mask on. Once I was away from the gas, a new panic came over me; I had dropped my M-16 rifle and it was nowhere in sight. I frantically searched for it, dashed back to the ravine; no rifle. Oh God, this was not a good situation.

Soon the exercise concluded and we were called back to formation. Do you remember why formations were held? To review how you screwed up. This was worse than throwing up in Basic. Everyone else had his weapon, except me. Good news was, it was not lost. The bad news was, the platoon sergeant standing in front of the formation had it and he did not look happy. Actually he probably was smiling inside, sadists that they are.

Upon calling out "*Who lost their weapon?*", the platoon sergeant glared at me as I raised my hand. I do not remember his words at the time, only the ridiculous punishment he quickly doled out.

He ordered me to climb a nearby tree, and sit in a fork maybe ten or twelve feet up. Then at the top of my freshly seared lungs, I had to yell out "*Croak, croak, croak I'm a tree frog – sorry I lost my weapon*" over and over again.

What benefit or lesson this was to surviving battles in Vietnam escaped me. FUBAR.

~ ~ ~

I should be grateful this was not the Roman Army. Discipline is the key to success in military units, in training just as in battle. Early Roman practices include a tactic called decimate. An early report from the fifth century B.C., tells of a Roman commander killing every tenth man in his army as punishment for desertion during a battle.

Hence the word decimate is forged from the decimal root of 'deci.' Roman Armies practiced this tactic of 'leadership by fear' frequently, as did other military forces throughout history. Fear is a large part of basic and advanced training. FUBAR.

~ ~ ~

The Vietnam War escalated quicker than the Army could produce experienced leaders. With the draft supplying much of the manpower, few soldiers were re-enlisting after their two-year commitment. This consequence of a conscripted army, particularly during an unpopular war, was a shortfall of experienced soldiers with leadership skills to command troops even at the squad level.

One of the solutions was to 'create' experienced Non-Commissioned Officers (sergeants) from new recruits. A new leadership program at Fort Benning, Georgia, modeled after the longer Officer Candidate School, manufactured sergeants in three months; one of the less than flattering names for these new NCOs was '90-Day Wonders.' This program recruited the best soldiers from basic and advanced training, and then groomed them to be squad leaders, the main role of E-5 sergeants.

After they graduated from the NCO program, these 'Shake and Bakes', another nickname for them, were assigned back into an AIT class as squad leaders to gain some actual leadership skills. Many times, they returned to the same place they were just three months earlier as trainees. The newly minted NCOs would stay in this role for one training cycle, eight weeks, and then shipped to Vietnam with their leadership skills finely tuned - hopefully.

This was a grueling scenario for fresh young recruits. Moreover, they received less than top-notch respect. Their superiors still considered them woefully inexperienced, and the trainees they were in charge of had few compelling reasons to follow their advice since it was all from school. They possessed no combat experience.

In their defense, it was an admirable path they took, and by all accounts they were outstanding leaders in battle.

We still made fun of them however, and held them only in slightly better regard than our own inexperienced selves. There was a different look in the eye of an instructor who had been to Vietnam and those that had not seen the gates of hell.

A pair of these new sergeants were close friends and each had their wives with them. One couple had just bought a new puppy and they all went on a picnic to a nearby lake: Alligator Lake. Why do you suppose it was called Alligator Lake?

Without offering any gruesome details, they lost the new puppy to an alligator. That was FUBAR itself. I guess *alligator avoidance* was not taught in Shake and Bake school. However, their inexperience with 200 recruits made the situation even worse; they never should have told the story. Now, behind their backs, around the barrack's corner, somewhere out of sight; someone was always mimicking the cries of a distressed, yelping dog. FUBAR.

Alligator Lake still exists at Fort Polk, it is even a popular family playground with pavilions and picnic areas. I assume they removed any alligators from the lake; especially since it is advertised as a family friendly place. It is not clear if dogs are allowed.

~~~

The last life-changing event at the end of Advanced Individual Training was receiving orders for deployment to a real duty station. In this case it was a sure bet that the lyrics "next stop is Vietnam" from 60s protest-rock band *Country Joe and The Fish* would be the #1 top play of the day.

We lined up again at an outdoor bulletin board, where a couple dozen pages were pinned side by side, each holding as many names as would fit on a page, in alphabetical order. Your entry was three or four lines line long because of all the jargon, acronyms, and codes required for orders.

The key you were looking for was the location of your next assignment. If it was Oakland, California or Fort Lewis, Washington, you were on your way to Southeast Asia. These were the outbound replacement and transit centers with nothing but plane tickets for either Tan Son Nhut Air Base outside of Saigon or Cam Ranh Air Base in the middle of Vietnam, both of these were large logistic bases with in-country transit centers.

In this case there was no hope, everyone went to Vietnam. However a couple of questions began circulating immediately: What is Fort Sill? And, where in the hell is it? Surprisingly, this was the destination for about 30 fresh infantry graduates, including me.

To add confusion and bewilderment, our orders included assignment to something called the 82nd USA Missile Detachment. It seemed like some kind of mistake, the Army is rife with those. Two weeks before I was watch-
~~~

ing a North Vietnamese enemy sapper (a highly trained stealth-like fighter who could penetrate defenses), who had defected to our side, demonstrate how he could sneak through a barbed-wire protected perimeter fence and throw a bag of TNT into a bunker without detection. Now I was going to Oklahoma? That lesson will come in handy I am sure.

Eventually we got a some explanation, but just a little. We were to be trained as security guards for a new missile unit being formed. Oh, and the missiles had nuclear warheads. This turn of events was interesting – in a couple ways for me.

Going to Oklahoma instead of Vietnam was the obvious surprise. However, remember the reasons for getting married? I skipped details of this part of my story up until now, but sometime between receiving my draft notice and this day of infantry graduation, I became engaged. Yes, to be married. In one week.

I would be lying and in denial if I did not admit to some regret right then of that decision. I was not going to die in the very near future, and I was only nineteen years old. FUBAR.

~ ~ ~

Chapter 5 – No Longer a Trainee

With responsibility for securing nuclear warheads, everyone was required to be in top notch physical condition. Our platoon leader previously destroyed his knee from jumping off a tank, and wore a metal leg brace.

~~~

Thirty days after finishing infantry training at Fort Polk, I reported to Fort Sill, Oklahoma, which is southwest of Oklahoma City about 70 miles along I-44. I was no longer a recruit, trainee, or low life embarrassment to humanity. Off to a real duty station where they do not treat you like scum – or at least pond scum. Maybe bathtub scum which is a step up in the scum hierarchy.

This was indeed a promotion in living conditions, not luxury, but on a relative scale, it was mid-market. No more barracks with 50 men to a room, we had two and four-man rooms. The bathrooms actually were equipped with sinks and urinals instead of galvanized World War II era troughs. These accommodations were actually better than some apartments I previously lived in.

Surprisingly, the food was phenomenal, grading on the curve against the mess halls at Fort Polk. These cooks liked their jobs, or so it seemed. And best of all, there were no dying cockroaches, no croaking tree frogs, and no overnight training exercises that included sleeping with snakes.

~~~

Fort Sill, Oklahoma is the Army's artillery training center; it is a good place for that role since the area has a lot of open space where you can fire a 155mm howitzer shell out a few miles and not hit anything important. In addition to the thunderous canons and howitzers, the Army also had a bunch of missiles. These were big missiles; many people would associate this type and size with Air Force operations. There was probably a high-ranking scuffle in the Pentagon that resulted in giving the Army missiles. Not only did the Army have missiles, they were nuclear tipped, a rather curious description. Somehow, 'if only the tip is a nuclear warhead", then it is not so deadly.

Nearly all military branches give names to their weapons in addition to a formal description. These missiles were named in honor of General John J. Pershing, the commander of all the U.S. forces in World War I. Hence, we knew them as a Pershing I missile. They were tactical missiles, that is their range was relatively short, about 400 kilometers or 240 miles. The Pershing's mission was to thwart advancing Soviet armor units should they start rolling across Western Europe. During the Cold War, the Soviet bloc enjoyed a tremendous numerical advantage over U.S. and NATO forces, particularly in armor divisions. The only equalizer western forces could deploy was nuclear weapons, each with a relatively small explosive yield, but enough to take out a division of tanks.

Nevertheless, they were nuclear, and luckily, we never found out how that would have escalated any conflict.

The Pershing Missile System was portable; the missile, warhead, power generator, and communications module each mounted on individual tracked vehicles built upon the chassis of Vietnam era armored personnel carriers. Four Army battalions in Germany were equipped with the missile systems and two German Air Force wings.

Seventy five infantrymen, including myself, and approximately 50 missile, logistical, and administrative personnel were training for a new missile detachment that would soon deploy to a German Luftwaffe (Air Force) base at Lagerlechfeld, Germany.

So why were infantry troops needed in a missile detachment? It seems that international and United State law at the time prohibited Germany from possessing nuclear weapons. This was only 25 years after the conclusion of

World War II. However, the German military was allowed possession of the missile launchers. The concept was that an American unit, the 82nd USA Missile Detachment which I had just became a member of, would be in control of the warheads. The infantry platoon was trained to secure the warheads and learn how to actually destroy them if needed due to being overrun by the enemy. The U.S. Army missile experts trained to mount the warheads on missiles that the Germans owned and controlled.

We, the U.S. missile detachment, were to be based in the middle of a German Luftwaffe base, guarding the warheads in a bunker, surrounded by German guards and German Shepherds as big as small horses. Should war have occurred with the Soviet Bloc, and firing the missiles was ordered, a convoy of Americans, with the warheads, and Germans with the missile components would leave the Luftwaffe base and drive to a remote location, mate the warhead with the missile and send it 'down-range' as they say. If all went according to plans, a nuclear explosion would occur on top of advancing Soviet tanks.

Now all of this missile, war planning, and history coverage is for background. I am not making comment or judgment of the overall strategy, battle concepts, or war planning. In addition, nothing here is secret nor am I divulging crucial information. Goggling any of the terms and names in the book will give additional detail well beyond the words I am writing.

The good stuff I want to share with you needs to be understood in the context of the sensitive and challenging environment at the time.

~~~

Enormous responsibility comes with custodial ownership of nuclear warheads; high quality personnel and training standards were demanded or, we should say specified. A couple of the requirements were that soldiers could not have a physical profile, and would need to pass scrutiny for a secret security clearance. Profile is used here in military slang; a physical profile is a detailed specification of physical qualifications. However, if someone has
~~~

a profile, they are usually restricted from some rigorous activities due to a physical ailment or disability. Soldiers in the missile detachment were not to have any profiles.

The top enlisted man in the infantry platoon was a sergeant first class, which means plenty of experience, a professional soldier, and one that could lead under any circumstances. It would seem that he should be an infantryman also, since he was the infantry platoon leader.

Our infantry leader was a tank soldier; his MOS was armor. In addition, at some point in his career, as the story went, he jumped off a tank and injured his knee or leg, requiring that he wear a metal leg brace at all times. As an infantry leader, it is a given that swift movement on foot may be a requirement. This was only the first of many questionable personnel assignments to our elite unit. FUBAR.

~~~

Besides requiring superior or at least above average physical condition, one would think a nuclear warhead custodial unit would have exemplary leadership at the first sergeant position, the top enlisted man in the unit. While ours was not an intellectual giant, he was crafty like a fox. However, foxes are also sneaky thieves.

My good friend and roommate for a while, Melvin, was also an infantryman plucked from the jaws of Vietnam. He was working on a Ph.D. in psychology at Oklahoma State when his student draft deferment expired. Recognizing that he was someone that could get things done, the first sergeant snagged Melvin for the lead company clerk position, overshadowing the trained soldier who actually went to the clerk MOS school.

Mel, as he was called, wrote most of the first sergeant's reports, and ran the unit in a M*A*S*H-like Radar O'Reilly fashion. One day while reading a report Melvin had written, the first sergeant blurted out from his office to the reception area where Mel sat, "Brook'ins, what does this here rapport mean?" The first sergeant did not know the meaning or pronunciation
~~~

of rapport, it came out 'raport.' Oh, and Melvin's name was not Brook'ins either, it was Brookings. The first sergeant was as stereotyped as you can get.

As for the crazy like a fox attribute, the first sergeant was a wheeler-dealer. That in itself is not bad, it takes some creative operations to get things done in the Army. He crossed the line when he started selling cars to his own enlisted men.

Once we arrived in Germany, any soldier could buy a car. Of course, most enlisted men had little money, and were frequently ripped-off by unscrupulous German used car dealers. (Author's note: I know this first hand, FUBAR). Well, the first sergeant wanted in on that business. Since he was a veteran of several tours in Germany, he knew the ins and outs of buying cars on the local economy. He would buy cheap used cars and then resell them to his own troops, pocketing a profit. This was a severe breach of protocol, abuse of position, and probably illegal. FUBAR.

In all due respect to the auto-dealer first sergeant, he died in 2008 and is buried in the Fort Bliss National Cemetery.

~ ~ ~

To continue with our not best and brightest theme, our quartermaster sergeant was at the top of the list. A quartermaster is in charge of everything that is purchased and issued to the men or used in the unit's operation. From toilet paper to uniforms, boots, and ammunition, he has the power. Frequently this is the position of wheeling and dealing, at least it is portrayed that way in most combat movies.

Our quartermaster sergeant had the distinction of serving the most years in the Army at the E-6 staff sergeant pay grade. This meant he was not very promotable. During the Vietnam era, a good soldier could rise to E-6 in four years, E-7 probably in less than 10. This sergeant had about 20 years in the Army, including service in the Korean War. He could have been a poster-boy for unpromotable men. This was another example of less-than-top-notch soldiers shoveled into a critical unit. FUBAR.

However, our quartermaster did have a very capable assistant, a Specialist 4th Class that was a good soldier and person, though he is another piece of our strange personnel puzzle. He was Canadian, yet the U.S. Army drafted him. Many may not realize that if they are a legal resident, such as a green card holder, or even an undocumented alien, they are required to register with the Selective Service System. This is true to this day, and they are subject to the draft. Drafting of non-citizens during the Vietnam War was quite common.

Since Albe, his first name, was Canadian, it is admirable that he accepted the draft to stay in the United States. Avoiding the draft by returning to Canada may not have been illegal, though it probably would have affected any future chance of returning and staying in the U.S. The interesting part is how his status related to our detachment is that we were all required to have secret security clearances; and you would think everyone would have to be a U.S. citizen in a nuclear-armed unit. FUBAR.

The above personnel reviews are not an indictment of any of the people involved. Most were career soldiers who found a good home, and the Army needed career soldiers. What stands out is the scaling problem the Army had at the time, and the consequences endured. It seemed that our newly-formed, critical unit did not receive top-level troops in many positions. Moreover, it almost appeared that in some cases the unit was a dumping ground for misfits or challenged individuals.

~~~

It was during this time that I finally learned how to stay invisible and not get into trouble. However, one day I hesitantly volunteered for something that would get me out of KP. Kitchen Police, also known as KP, is where you are the cooks' slave-for-a-day. How the term kitchen police was conjured up is a mystery; you definitely do not have any police powers. If you are not peeling potatoes you are most likely washing pots and pan, or mopping floors. Most any duty is better than KP.
~~~

The first sergeant was asking around to see if anyone could type. My friend Mel was already the de facto company clerk, and we had a trained one, I was not sure why the top sergeant needed another typist.

I broke the soldier's cardinal rule by volunteering. To rule out any suspense, it worked out well, and produced an opportunity that leads to many more stories. Poor planning by the Army gave me a break.

Pentagon planners define logistic requirements, equipment needs, and personnel staffing for military organizations of all sizes. They create a Table of Organization and Equipment document for every type of unit in the service. There are many variations, I will only cover the concept here – it is a boring subject.

A TOE (or T/E or BTOE, ad nauseam) describes what a military unit looks like: Staffing, numbers and types of weapons, office equipment, trucks; everything needed for full operation. The people who make up TOEs have probably never actually been in an operational unit. Ok, maybe not a fair or accurate statement, but the message is valid.

There is never enough equipment or staff to conduct the mission correctly. My friend Melvin's new position as company clerk is a prime example. The TOE specified one clerk, which the first sergeant and the commanding officer deemed not enough. Taking one infantryman from the security squad and making him a clerk would not compromise the mission. He was still listed as an infantryman and no rules are officially broken. I volunteered to solve a similar problem; one that I did not know existed. I became the detachment's Training NCO.

Every Army unit down to company level has a Training NCO. The job of the Training NCO is to schedule, manage, and track periodic training of topics such as first aid and weapons qualifications. The company clerk usually performs most of the administrative work such as updating each soldiers training record, while the Training NCO has mostly managerial responsibilities. A Training Officer will also help navigate requests and arrange for difficult logistics. The job of Training NCO is not that difficult in a standard infantry company, and is usually assigned to one of the platoon sergeants.

In the case of forming and training a new missile detachment with nuclear responsibilities, managing the training regimen, keeping records, and preparing for inspections and certifications is a full time job. A Training NCO was not part of the TOE for the 82nd Missile Detachment.

I knew nothing of managing training, barely knew the ways of the Army, I was only a PFC (Private First Class), and most definitely knew nothing of missiles and nuclear warheads. My only qualification was that I could type.

To be clear, I was not in charge of making things happen, I reported to a first lieutenant who had been to missile school, and the officer staff was fortunate to include a senior Warrant Officer who was known and respected around the world for his knowledge on the Pershing Missile system. I was a glorified clerk. However, a good position just the same. It kept me out of KP, I had no weekend or night guard duties, enjoyed my own office, and eventually gained a good deal of respect. Finally, a task that was not FUBAR.

~ ~ ~

The strange thing about a missile detachment is the number of officers included in the TOE. A combat infantry company may have about 125 to 150 soldiers with six or seven officers, a commanding officer (captain), an executive officer (first lieutenant) and a platoon leader (second lieutenants) for each platoon. Head count can vary partially due to the number of platoons, which could be four or five.

Our detachment consisted of 126 men: 11 officers and 115 enlisted men. The commanding officer and executive officer were normal positions, plus an infantry officer lead the security team. The remaining lieutenants were artillery officers with some missile training.

Most of the artillery officers were green as the infantrymen; they had just come out of OCS (Officer Candidate School), so they were as dumb as we were but were not supposed to act like it. Like many of the infantrymen, they were glad for an assignment other than Vietnam where they likely would be camped on a remote firebase for a month or two directing artillery fire.

One officer, the first lieutenant that I reported to, went to great lengths to avoid Southeast Asia. To digress a little, remember the story in the induction center about signing up for extra years for a comfortable or better job? Well, officers can do the same thing. The story went that when he came out of OCS and was assigned to be an artillery officer, he could avoid the barren Vietnam mountain tops if he extended for missile officer training. The catch required, as you remember, more time devoted to the Army for such privilege. In this case, I believe it was four more years on top of his current commitment.

The system can also screw officers. He believed by agreeing to an extended commitment his MOS would be changed from field artillery to one of missile MOS. Your MOS classification is crucial to the determining which duty stations you can be assigned. Just about every MOS was involved in Vietnam. However, there were no missiles of this type in Vietnam and no tactical nuclear weapons. 1st Lt B., as I will call him, reasoned that MOS exclusion from the area was a sure way to avoid deployment to hell.

Lt. B. either did not read the fine print (or the large print), was lied to by the enlistment officer (imagine that), or something just fell through the cracks. His MOS was not changed from artillery; therefore, he was still subject to Vietnam deployment at any time manpower needs exceeded the number of available artillery officers.

He spent hours writing letters requesting his MOS classification be changed to missiles. I lost track of him, his enlistment went for many years after I left the Army, which could still have put him in Vietnam. FUBAR, but funny.

~ ~ ~

The time spent at Fort Sill, Oklahoma was mostly routine and un-eventful. I hitchhiked nearly every weekend either to Fort Smith, Arkansas (not a military base) or to Wichita, Kansas which was my original hometown where I still had many friends. I would meet my new wife at my parent's home in Ft. Smith; she would take a bus from Pine Bluff, Arkansas to Ft. Smith, which was about a mid-point between us.

Times were different then. This was 1969 and soldiers in the mid-west were mostly afforded respect. I would put on my dress uniform, and ride to Oklahoma City with a fellow soldier from that city. He would let me out on the freeway, and then I would easily hitchhike along Interstate 40 to Ft. Smith or Interstate 35 to Wichita. I never spent more than a few minutes with my thumb out.

Once I almost caused a wreck on the interstate. While thumbing a ride on I-35 north of Oklahoma City, a couple of cute girls were trying to pull over to give me a lift north towards Wichita; at the same time a man in another car also tried to pull over. Their cars almost tangled. The girls drove on and I had to ride with some boring guy for a couple of hundred miles. I still wonder about those two girls. FUBAR.

~~~
~~~

Chapter 6 – FUBAR in Deutschland

The disbursement officer told us, very literally, that he did not care if we threw the ammunition into a river once we drew it from his group, we just had to requisition and receive it.

~ ~ ~

With nearly six months of training complete at Fort Sill, the Army decided it was time to deploy the detachment to its operational base in southern Germany. This was exciting. I had not traveled outside the U.S. before, and our family vacations as a kid were not particularly joyful. Germany would be a great experience; living in the beautiful Bavarian region with easy access to world famous beers was a dream come true for many teenaged solders.

After 22 days leave, the complete unit would report to McGuire Air Force Base in New Jersey by 3:00 P.M on November 26, 1969, the day before Thanksgiving. A charter flight was ready to fly us overnight to Frankfurt, Germany. Military planes were in such short supply due to Vietnam requirements, that independent charter companies transported most U.S. personnel between countries, even to Vietnam. We landed at Rhein Main Air Base near Frankfurt, Germany on Thanksgiving Day 1969. Holidays are not sacred in the military.

~ ~ ~

Duty in Germany produced many great experiences and stories, mostly good, yet also several surprising FUBAR moments.

Our commanding officer was an artillery captain with extensive combat experience; a North Vietnamese Army unit overran a remote firebase under his command. During the battle, one of his men was repeatedly wounded, yet continued to direct fire and rallied his men to repel the enemy.

Our CO, Captain M., worked tirelessly for two years pushing forward the process of a Congressional Medal of Honor award for the brave sergeant in his battery.

President Nixon presented SFC Webster Anderson the Congressional Medal of Honor on November 24, 1969, two days before our deployment to Germany. The story within the detachment was that our CO, Captain M., was very disappointed that he did not receive an invitation to the award ceremony at White House. FUBAR.

Captain M. was the most respected officer in the detachment, deservedly so. He was career officer, already awarded numerous purple hearts and medals of valor. In addition, his leadership skills and professional stature was not questioned. This would be important, as many on his staff in Germany probably tested his patience and management kills.

~~~

Our living conditions were very nice, again grading on the curve. We were on a German air base, and all the barracks were probably early 1960s German construction. We lived in the same conditions as the German troops; two-man rooms, steam radiator heat, and a large common shower and latrine in each wing of a low-rise one-story building. My understanding is the German base command was always complaining about our heating and hot water bills.

For those of you who are not aware of customs in Germany and throughout much of Europe at the time, room temperatures in the winter were quite chilly, and showers or baths were generally enjoyed but once a week. This was not just a military tradition, it was the normal way of life for the general populous. European countries have always endured much higher energy prices than the United States, and conserving energy for heating and hot water on-demand has always been a priority. The Americans were viewed as very spoiled if not wasteful.

My soldier buddies and I did put the winter weather to good use however.
~~~

We would chill our wine right outside our window. I was not much of a wine drinker at 19 years of age, knew little about wine or even cared much for it. With frigid temperatures in the winter, there was little to do after the day's work; some of us started playing cards every evening, usually hearts.

The first week or two one of us would go to the on-base German PX and get a bottle of wine. The four of us would make the single bottle last all evening while playing cards until lights-out. After a couple of weeks, we decided that two bottles of wine were necessary to make it comfortably through the evening. This was white wine, so putting it in the snow on a ledge outside a window functioned as our 'wine bucket.'

At the beginning of this tradition, besides not being a wine drinker, I was not a good card player either. However, practice makes perfect as I was getting quite proficient at cards and drinking wine. Within a month or so, it was still only January and brutally cold, each of us had our own bottle of wine for the evening. (Good thing we were not in possession of those warheads yet.) Besides me, our foursome included the head company clerk and the two cryptographers who would decipher orders to attack, or retreat, or whatever a secret message would say. I am sure in secret-code training classes the brainiacs were tested for speed and accuracy in message deciphering. However, were they tested under duress or influence of a nice sweet Gewürztraminer? FUBAR.

~~~

Crypto (short for cryptography) people are a little different than most. Though, from spending my entire after-army life in the computing field, I have concluded that computer programmers have pretty much the same skill and mindset as secret-code people. The TOE allotment for our detachment was for two trained cryptographers. Once we would become certified to handle the nuclear warheads, one of them was to be on duty, usually in a sound-proof, electronically shielded room 24 hours a day, seven days a week; since you could never predict when the Russians might attack. Simple math calculations show us that this would be a crappy job with just two of
~~~

them; more than two cryptographers were needed. This shortage of crypto personnel would create the need to grab another infantryman that can type to fill the void.

Luckily, Specialist David B. had been hanging out with the crypto guys, was fascinated by the job, and he wanted in. I bet he became a programmer in civilian life. David B. was a perfect fit, he would learn via on-the-job training how to decode messages and save us all from unknown annihilations. He even lobbied the first sergeant to submit an MOS change for him; from infantry to crypto. Everybody wants to change their MOS. Unfortunately, since there was no slot in the TOE for a third cryptographer, that was never going to happen. He whined about it so much that his nickname became 'Slot', because he did not have one. Nicknames in the Army eventually become your near-real name, yet he relished it. So 'Slot' it was. FUBAR.

Staying on the crypto subject, the two officially trained cryptographers were special cases themselves. They both had college degrees and both were schoolteachers. Not sure how they came to be cryptographers, probably fell for that "sign up for four more years and I'll send you to school" routine. To many recruits this sounded much better than grunting through jungles or hiking up mountains with 80 pounds of gear on your back as an infantryman; not nearly as dangerous either, or so they thought.

What the cryptographers did not know at signing time, was they were in very high demand in Vietnam. There were still secret messages that needed coding and decoding, even though the message was not warning of 10,000 Russian tanks storming our way. What is worse, there was actually a bounty on the heads of cryptographers in Vietnam, placed by the Viet Cong or the North Vietnam Army. That is crypto lore anyway.

This potential danger was not something that one of the cryptographers expected, especially since he received orders for Southeast Asia right out of cryptography school. Obviously racked with fear and conflict of duty, he broke his own arm while on leave before deployment, thinking it would at least delay his deployment. I am not sure of the story he fabricated, but it worked. After convalescing at a temporary stateside base, this brave young man was assigned to our detachment. Another example of only the finest assigned to sensitive positions. FUBAR.

~ ~ ~

Most soldiers love being stationed in Germany. Easy access to great food, good beer and wine, and enchanting sights previously only seen in books or movies. Everywhere you drove, another 500-year-old majestic castle sat near the highway. Our base was just north of the Alps in Bavaria only 70 miles from the famous Neuschwanstein Castle near Füssen; Walt Disney reportedly fashioned the iconic Disneyland castle from this magnificent fortress. Captured by its beauty and massive construction, I visited it several times.

Most of us learned to speak a little German, at least enough to order beer and wienerschnitzel at the local gasthaus. Speaking some German would gain you a little respect with the locals, even if your language skills were lacking. All local young men and women spoke English, as it was required curriculum in public schools.

I was very proud of one German language skill I honed – cursing. I even received a compliment on it by a German soldier. One day I was in the German PX which also held a bar, restaurant, and a game room. While playing a pinball machine, a dedicated pastime from my early youth, I somehow missed a flipper shot or otherwise lost the ball and game. With perfect diction, I blurted out 'scheisse', the universal reaction to losing a game. If you do not speak German or have not made the translation yet, that is the German word for 'shit.' This is required language for pinball machine players.

A German soldier was watching me play; this is a great way to build international relations. I was in civilian clothes, yet he assumed I was an American anyway, usually not that hard to distinguish from a German soldier. Surprisingly he turned to me and inquisitively asked "You are American right?" I responded yes, in English since I did not have conversational German-language skills. Much to my delight, he then said "you pronounced 'scheisse' in perfect German." That made my day and obviously, it was a proud moment that has stuck with me all these years. A soldier's life was not all FUBAR.

~~~

Our stay in Germany was not only about good food and beer. We had a mission. Nearly a year of additional training was required in Germany before operational certification could be granted, and only then would nuclear warheads be given to the unit. A lot of classes, exercises, paperwork, tests, and many inspections were required for certification.

Since we were on a German Luftwaffe base, we did not have access to many training facilities that were available on American infantry bases. The 1st Infantry Division, or Big Red One, was based north of us about twenty miles in Augsburg, though we had no direct reporting or connection to them. Except, if we were in trouble, big trouble. If our base, and the warheads, were under attack, our crypto experts would alert the Big Red One who would then charge down Highway 17 to our rescue. I am sure chaos would have ensued.

As part of our TOE we received a shipment of brand new M-16A1 rifles, still in original crates and factory packing grease. One of the required training exercises was for everyone to re-qualify on the M-16. We needed to find a practice range and then pass standardized marksmanship tests to become certified.

Being the Training NCO, I was now a Specialist 4th Class, it was my job to schedule the practice and tests. I was not on my own, Lt B. took the lead on difficult challenges, plus as an enlisted man I did not have authority to cut through a lot of red tape.

A couple of officer-level challenges cropped up right away. We had no ammunition and no rifle range. The ammunition was not that difficult to acquire. Just fill out a lot of paperwork for our headquarters in Ulm about seventy miles away, and then drive to Ulm and pick up the M-16 rounds.
~~~

The bigger problem was lack of a rifle range. There was not one on the German air base, and most U.S. training sites were quite a distance from us. Luckily, Lt. B, the training officer trying to change his MOS, took the lead in arranging all of this. First, get the ammunition.

After filling out supply requisitions in triplicate, sending them along with forms that justified our request, we were authorized to pick up the ammunition from our headquarters in Ulm. As part of the arrangement with our deployment on a German base, they supplied us with many services. One service was transportation, since our TOE did not give us trucks, jeeps, or anything with wheels. So, we arranged for a German driver and Volkswagen bus to take the lieutenant and me to Ulm and pick up the ammo. We still did not have a rifle range to use, but one thing at a time.

Lt. B. and I met with an officer in Ulm, a major I believe, for introductions and instructions on how to draw the ammunition. As a function of compartmental responsibilities in the Army, this officer and his office were only responsible for giving us the ammo. Lt. B. explained that we did not have a rifle range to use, and asked if the major could offer any suggestions or arrange the use of one. His answer came under the category of 'not my job.' He told us, very literally, that he did not care if we 'threw the ammunition into a river' once we received it from his group. That was probably a standard answer for many requests he received, since the Danube River runs through Ulm. His job was to fill the requisition, account for the ammunition in inventory reports up to the time he turned it over to us, and make sure we drove away with it. FUBAR.

Lt. B. did deserve respect for his diligence. He ensured that we did everything correct, including becoming proficient and certified with our new rifles, and that we would not throw the ammo into a river. Somehow, he found a German rifle range and scheduled a series of practices and qualification exercises. The unit's M-16 rifle qualification records were true and accurate.

~ ~ ~

Not all unit-training records would be 100% accurate. I was responsible for maintaining records of each individual in the detachment showing when they competed training on everything from first aid, physical fitness, and security procedures, to warhead handling; literally dozens of disciplines had to be scheduled for training. Classes and training were conducted at least twice, the second being a make-up class for those that could not attend the first.

Separately, I created and maintained a master training schedule. These schedules were reproduced by mimeograph machines and sent to various commands, maybe even back to the Pentagon where they would probably be filed and not read.

It was just as important that the files, reports, and schedules exist, and in perfect order, as it was for the activity to be conducted. I considered a good portion of my job to include magic – dates, records, and activities just appeared.

Now before we call a congressional oversight committee to investigate, 95% of the records were accurate. It was usually make-up classes and events that needed a little adjusting. Even if you scheduled a make-up class for missing students from the first one, some still might miss the make-up class. It was impossible to catch everyone, every time for a scheduled class; some teaching might be on-the-job training or an informal one-on-one session with the instructor.

~ ~ ~

So why were we concerned about perfect paper work? Inspections. Once we were well into training for certification, a team of high-ranking inspectors would arrive at least once a month for detailed reviews of our progress – especially progress reflected in our training records.

Rank has a strange way of transforming even the most confident and hardened soldier from the pillar of stoic strength their troops know them as, to near-quivering subordinates in front of higher ranking officers. The greater rank differential, the greater the quivering. As a young draftee new to military protocol, it was interesting and educational to watch the dynamics of command at work.

The inspectors were not only from our headquarters but also from NATO headquarters. These teams included colonels and generals, as well as their aids.

To counter much of the chaos portrayed in this book at the individual level, these officers took their job and responsibility very serious. It was obvious they would make sure our new detachment was worthy of receiving nuclear responsibility. It was not always fun having them invade our space, as they would scour my records thoroughly.

If discrepancies were uncovered, the infractions were included in a report given to our commanding officer. A major commanded our detachment (recently promoted from captain) and our technical officer was one of the highest ranking and seasoned Warrant Officers in the Army. Yet, they became nervous wrecks on inspection days.

When they were admonished by the inspection team, you can bet once the inspectors left, the chewing would continue down the chain of command to this lowly enlisted man. From our commanding officer to the first lieutenant to me. Then the first sergeant would come in for a blind side body-blow. FUBAR.

~ ~ ~

We had our share of misfits in the infantry ranks. One in particular was from the rough streets in New York. Not one for discipline, he was frequently in trouble form missing formations, or late returning from leave.

One night he returned to the base quite drunk. This was a German base, and the German version of MPs (Military Police), staffed the guard station at the entrance. Our man either was belligerent or forgot his ID, as the confrontation proceeded to where he pulled a switchblade knife on the guards.

He must have forgotten the cardinal rule that you do not bring a knife to a gunfight. The German guards were well armed with Luger automatic pistols. No one was harmed, but our man was arrested. This was another example of exemplary troops in our unit. FUBAR.

However, his saga continued. Several months later, he received orders for Vietnam. Apparently, that was not on his agenda, so he went AWOL. The story goes that he pimped for several prostitutes in Augsburg. I never did hear if they caught him, though he was in a foreign country with only military ID and no passport.

Chapter 7 – Oops, Never Mind

The winter weather finally gave up and let us enjoy the new grass and flowers that winter snow nourished. Springtime in Germany is a time for renewal. It is a time for long strolls along promenades and country roads. However, the most exciting springtime events are the early beer festivals.

Most everyone is familiar with Oktoberfest, which actually begins in September and ends the first week of October. This overly-commercialized celebration is only one of hundreds if not thousands of beer festivals held throughout Germany. The country always supported a large number of breweries, nearly every town included a local brewery. Even today, the German Beer Institute list 1,200 German brewers. Most of these are congregated in Bavaria and southern part of the country, and along the Czech Republic border. Luckily, I was in Bavaria.

Beginning in the spring, local breweries would setup big circus-like tents, entertain everyone with local oompah band with musicians wearing lederhosen, and tap fresh kegs of beer that were up to six feet in diameter. It is an astonishing sight to see a keg that large, on its side in the back of a large truck. They did not unload these uber kegs; just tapped them in place.

Every weekend a festival could be found within an hour or two drive, or closer. The festivals continue through spring, summer and conclude in the fall.

~ ~ ~

For many of us, these times would be the swan song for our German adventure. Do you remember from my introduction about President Nixon's promise to reduce forces in Vietnam and reduce the draft? He did indeed begin action on election promises shortly after his inauguration. There is however, a lesson to be learned about political promises and math.

Draft quotas began to plummet during 1970; these announcements provided great press stories for a country tired of watching the draft absorb the youngest and now the brightest as college deferments began to expire. In conjunction, planned troop reductions for Vietnam were proudly highlighted in press briefings and speeches.

The math was troubling however, even though the average American was not aware of an imbalance between troops still needed and the number of new recruits entering service. Draft quotas were cut faster than troop levels were reduced. This created a shortfall of replacement troops that needed to be filled from some source. What better place to find more troops than Germany?

U.S. troop levels in Germany were mostly spared during the mid-1960s. However, the dramatic Vietnam troop build-up during 1968 caused U.S. force draw-downs in Germany, and the subsequent draft reductions in 1970 triggered an even greater number of transfers to Southeast Asia from Germany. The Official Home Page of the U.S Army in Europe website carries the observation:

"In the late 1960s and early 1970s, the needs of the war in Vietnam reduced USAREUR's assigned strength, sometimes drastically."5

The Tet offensive (so named as it coincided with the Vietnamese New Year, Tet) in January 1968 with its coordinated attacks in nearly every South Vietnamese city was a wake-up call that the enemy could do whatever they wanted. Though the offensive was beat back, at very high cost on both sides, the follow-up response by the Johnson administration was to increase troop levels – and draft levels – throughout 1968.

President Johnson's election defeat in November 1968 would reverse the escalation, though not immediately.

~ ~ ~

By the summer of 1970, many troops stationed in Germany and throughout Europe were being sent to Vietnam. The Selective Service draft was not keeping up with combat replacement demand and infantrymen were especially vulnerable to the transfer.

The Army uses the term 'levy' to describe orders of transfer from one assignment to another. A levy can come down, you can be levied, you can receive a levy; there are any number ways the term is used.

No one in the 82nd USA Missile Detachment was even thinking about going to Vietnam. The average enlisted man probably was not aware of the trend that was happening elsewhere, I was not. Besides, most of us in the infantry were draftees and our discharge date was well less than a year away, most between six and eight months. Even the Army would not send us with that little time, especially since the clock was ticking and you always received 30 days leave before deployment to a combat zone.

As the 60s fictional television character Gomer Pyle would say, "Surprise, Surprise."

The middle of July delivered a levy to eight of us assigned to the infantry security platoon. I was still officially part of that platoon. Here is an example of how the MOS designation dictates your future. The levy was for eight men with the MOS of 11B from the 82nd. This was more than 10% of the security force. The names were chosen at a higher headquarters, no one in our detachment including the CO or first sergeant had any idea the transfers were coming or who was on the levy. The list also included my friend Mel, the company clerk who was still officially an infantryman.

Our orders specified we were to report to the replacement depot at Fort Lewis, Washington on August 18, 1970 after a 30-day leave. FUBAR

~ ~ ~

Besides the personal disruption and anxiety, it seems puzzling why the Army would go to such lengths to build up a special unit of this type, and invest the time and money for training, then jeopardize the mission by reducing the force. Especially, since this levy would only supply eight infantrymen to the vast pool of men going to Vietnam. The only rational thought is that troop cuts were blindly applied across-the-board regardless of the impact to particular units.

The transfers were so widespread that one of my friends returned to the United States sitting in first class of a commercial airliner because military and charter aircraft were in short supply.

The scramble to pack, return equipment, process out, and begin the trek to Vietnam was somewhat frantic, and certainly disruptive to the detachment. Other key people on the levy included two of the top infantrymen who were squad leaders, good friends of mine who I would be close to once we arrived in our new Southeast Asia station.

The commanding officer arranged for a nice farewell dinner for all of us; it was a strange farewell. FUBAR.

~~~

My friend Mel was a different sort than the average teenage draftee. Along with his studies for a Ph.D. in psychology, he embraced many of the thoughts on war that swarmed over campuses, and was not sure he would report for Vietnam duty. He even talked about going to Canada. Mel really was conflicted.

I never did see him after Germany, never exchanged any correspondence with him and have not been able to track him down, even with the power of the internet. However, I did hear from other soldiers from our Germany days, and the word was he did report for deployment to Vietnam. He became a squad leader in an infantry company, and was an excellent sergeant, some said he was real gung ho. You never can predict what people will do under adverse conditions. Mel, contact me.

~~~

The grand plans that someone in the Pentagon drew up for tactical nuclear weapons several years before was now cast aside, or at least shrunk. The Soviet threat took a back seat to the math of politics. NATO forces did not need, or could not afford, the 82nd USA Missile Detachment.

After the first one, two more levies were placed on the 82nd, one included two artillery officers. I actually talked with them as they were processing into the 101st. One, Lt. M., was a stand-up guy; a graduate of Ball State University. Naïve that I was, I had never heard of Ball State and thought that was a funny name for a school. The other lieutenant was all right, though I do not remember his name. They had a deer-in-the-headlights look about them upon arriving in Vietnam – I am sure we all did initially. Their world went from nice dinners in a German gasthaus and clean sheets, to commanding artillery batteries on remote mountaintop firebases eating C-Rations and sleeping on wood planks. FUBAR.

Back in Germany, the staff reductions hurt the detachment's ability to become certified for nuclear responsibilities. Besides, who was making the training records look good since I left?

Within six months, the 82nd USA Missile Detachment merged with a similar unit that was only ten miles away in Landsberg. Our commanding officer, Major M. returned to Vietnam in 1972 as an advisor to the South Vietnamese Army. It seems that everyone's experience with the 82nd would soon become a footnote of no consequence. FUBAR.

~ ~ ~

Strange situations would be normal procedure from this point on. Nearly 30 days of poignant, sometimes awkward dinners and visits, with family and friends was too many. I felt ready to get on with the show. However, I was not going to arrive early in Fort Lewis, that would be an overanxious plunge into the unknown.

I timed my flight to arrive at the replacement center on August 18th, just like the orders dictated. Now, we all thought it was strange that we would be sent to a combat zone with only six or seven months left in the service. The standard deployment was for one year. Yet, here we were. I met up with a couple of buddies from Germany, but not everyone. Some might have arrived early, others late.

The idea of deploying for six months was also silly to the Army. It takes a while to acclimate a soldier to a new environment, especially an inhospitable one, and once in Vietnam there were days and days of in-processing, assignment to an actual unit, getting to the new unit, and then a week or two of P-Training, short for Preparation Training or something like that. The last thing a first sergeant wanted to see was some green soldier that was just going to get some experience and become trustworthy and then the soldier's enlistment is up. I understand why first sergeants are usually so surly.

So, the Army's brilliant solution was not to prevent all this before it happened, but to draw a line in the sand at the last moment. Upon arriving in Fort Lewis, if a soldier was less than 180 days from discharge, he would NOT ship overseas. He would most likely stay in Ft. Lewis and pull some very boring duty like painting rocks along the side of the streets or pull a lot of KP duty.

However, any soldier with more than 180 days remaining would catch the next big bird to Southeast Asia. I had 184 days left in my two-year commitment. FUBAR.

~~~
~~~

Chapter 8 – Next Stop is Vietnam

One hundred and eighty days or less you did not go to Vietnam; I had one hundred eighty four.

~~~

Our Flying Tigers charter flight landed in Cam Rahn Bay, South Vietnam after whirlwind processing in Fort Lewis. Cam Rahn was one of the major shipping and transit points in the country. We were issued our jungle fatigues, combat boots, and all things suited for Southeast Asia, everything except weapons, which are issued by your assigned unit. The process reminded one of basic training, not the harassment, but the cattle-herding method of moving large numbers of troops through to their next destination.

It was at this point we would be assigned to a division or combat command. The Cam Rahn processing center supplied troops to combat units in the middle and northern areas of South Vietnam. This included the 5th Mechanized Infantry Division, the 101st Airborne Division (Airmobile), Americal Division, the 173rd Airborne Brigade, and the 4th Infantry Division.

On that day, seems like most of us went to the 101st, which was based near the cities of Hue and Phu Bai in the most northern sector of South Vietnam called I Corp.

The 101st was scattered among several camps, each within five to ten miles of each other; Camp Eagle was the division headquarters. Camp Campbell was the personnel processing center, Camp Evans, and a couple others.

It would take two plane rides to reach the 101st. One from Cam Rahn to Da Nang which was about 300 miles, then another short hop on a smaller plane to the Hue-Phu Bai airport.
~~~

The Da Nang airbase was enormous; the runway was over two miles long, capable of handling any aircraft. At one time this airbase handled more take-offs and landings than any airport in the world, most of those combat missions.

We took the final flight leg to Hue-Phu Bai on a small smaller, two engine transport, I believe it was a C-123.

Entering through a rear ramp into the belly of the plane, lucky passengers saw that the plane had no seats; only aluminum and nylon-web benches along each side, and down the middle. The seat belt was a strap that would stretch over a couple soldiers. Needless to say, there were no flight attendants.

By this time, the day was getting long and after strapping in, many of us tried to catch a nap. This would be difficult as once the engines were powered up and the plane started taxiing it rattled and shook as if it was going to fall apart any second. Also, the large propellers just feet from us created an thunderous roar. But, a good soldier learns to sleep under any conditions.

The flight distance was only 55 miles and after twenty or thirty minutes of nodding off and waking up, I asked the soldier next to me jokingly '*are we there yet?*' Well, the joke was on me, we had not even taken off. We taxied endlessly around the airbase. Yet, the shaking and noise seemed like we should have been airborne. My guess is that combat flights of bombers and gunships probably received takeoff priority. FUBAR.

~ ~ ~

Eventually we arrived at the Hue-Phu Bai airport that was a joint South Vietnamese Airport and U.S. Airbase. Taxiing around the Da Nang airport took longer than the flight.

With more group herding, we went through a series of 'get on the bus' and 'get off the bus' exercises as we made our way to Camp Eagle just several miles away. This would be our temporary home for a week or more.

We must have looked like the most naïve and scared tourist during the bus ride. This was our first look at a combat-engulfed country. It was hot, dirty, poor, and alien in every respect to most of us. Roadside shacks were cobbled together from discarded ammunition crates. Roof shingles were sometimes unrolled and flattened aluminum cans – you could recognize Coca-Cola, Budweiser, Fresca, and any number of other brands. Barefoot kids scampered over dirt floors. Blank looks stared at us; with no indication of like, dislike, friend or foe, hate or love. It was all a mystery.

I grew up in Kansas, spending time on my grandparent's farm and riding dirt bikes in open fields with friends. It was not uncommon to see coyotes hung over fences, killed to thin the packs and prevent raiding of farm animals. Their ears were cut off and turned in for bounty. I did not expect to see similar sights on this first bus trip – only they were human. The enemy. FUBAR.

These were displayed not for bounty but as an example or warning to others. I do not believe American soldiers did this, though it would be gullible to think it could not happen, but more likely South Vietnam soldiers were responsible for the acts.

~ ~ ~

Camp Eagle was the division headquarters and supported a week long indoctrination course for new in-country troops. Called 'P-Training', shortened from Preparation Training, this course was one of those "forget everything you learned in basic and advanced training, this is how we do it here" courses. We were back in the training mode, living twenty men to a barrack. The instructors covered everything from hygiene to weapons maintenance, tactics and patrol. Most of it was not classroom work but outside-the-wire real-life exercises.

The area was secure, so we conducted our first patrols outside the camp perimeter. The perimeter of a camp, firebase, or any defended area is ringed with barbed wire, coiled concertina or razor wire, and other obstructions that would keep the enemy out or at least slow them down. Anything beyond the perimeter is no-man's land. This patrol also included our first possession of live ammunition and loaded weapons in a real-world setting.

I think the veteran squad leaders were more scared than the new troops, as they were afraid one of us would panic, stumble, or otherwise shoot them accidently. They constantly warned us to keep our M-16s on safety, and frequently scanned the weapons to make sure we listened and obeyed.

These patrols were interesting and frightening; your mind wanders as to what could happen, nonetheless they were uneventful. Each evening we turned in our weapons, they were locked in a steel Conex container, the common name then for what we simply call shipping containers today. The staff did not trust us with deadly weapons at our side. Good call probably.

Other exercises include actual helicopter assault training. All through basic and advanced infantry training in Louisiana the extent of helicopter familiarization was jumping in and out of wood mockups.

Actual flight would be cool, and become one of those firsts that I will never forget. I was the last one to get on our Huey. The UH-1 helicopter was not very big, it carried only half a dozen infantrymen with packs and weapons, along with the two door gunners and two pilots. I was sitting on the edge of the floor, with one foot dangling outside the chopper; the doors were removed. After the first few minutes of clutching a structural piece of the Huey, particularly during a highly banked turn, it was actually a cool feeling. The image of the door gunner next to me, ready to unleash 600 rounds per minute from his belt-fed M-60 machine gun if needed, stuck my mind forever. He was an expressionless black soldier with his helmet painted in a one-color black design of Peter Fonda's flag helmet from the movie Easy Rider. This was the door gunner's way of rebelling.

We flew out past the edge of the camp to a landing zone, dismounted the chopper, and simulated an assault. The choppers took off and we formed a perimeter. Shortly the choppers returned and we flew back to camp where the trainers conducted a debriefing and critique.

After evening mess, supper that is, we went to the on-base enlisted men's club for a beer. Lest it sound like some luxury, it was a quickly assembled wooden building, I do think it had a floor, with a bar made from 2" x 12" planks, and a dozen or so broken-down tables and chairs. What a dump. The price was right however, a dime a beer.

We were having a nice cold one, when all of a sudden a not-too-distance thump and follow-up explosion rang out. We were all new replacements, however, it did not take a Rambo to understand that a mortar shell was lobbed our way. Almost everyone in the club stampeded out the door and sprinted to the nearest fortified bunkers that were scattered throughout the camp. These particular structures, you could call them bomb shelters, were built into the ground and covered with multiple layers of wood planks, sand bags, and a little concrete. They offered protection from most mortars rounds and maybe small caliber artillery shells.

Now I said almost everyone scrambled out of the club. The bar tenders and in-country veterans evidently stayed calmly put. I did not realize this until after the group panic attack. They knew the mortar round actually landed a safe distance away, and from experience knew it was probably a single harassing round. Therefore, they never left the building.

They simply went around the club and collected all the cigarettes and lighters left on the table by the newbies. These crafty veterans probably never bought their own cigarettes and enjoyed many laughs grading the new bunch of replacements. My loss included an engraved Zippo lighter I bought at Fort Polk and proudly used for more than a year. FUBAR.

~ ~ ~

Chapter 9 – New Unit, Same FUBAR

The jeep driver said to the kid, "Give me three vials of coke."

~ ~ ~

P-Training would end with no drama except for my lost cigarette lighter. Now it was back-on-the-bus. None of us was assigned to our final unit yet, that would happen at the next stop, which was Camp Campbell. This camp contained the 101st administration center; a bustling place where every new replacement, enlisted men and officers, passed through, receiving assignment to their destination unit. Over 100 new replacements came through every day, sometimes as many as 300.

We were again herded to different buildings and lines, trudging around with all our worldly possessions in a duffle bag. The last stop was at the end of a building as long as a football field. In front of us were two doors, one in and one out, each separated by about 30 feet. We lined up in front of the 'In' door and waited until called by someone inside barking "next."

Besides our duffle bag, the most important possession was the famous 201 file. This file, a heavy folder with metal fasteners holding papers in place, contained a soldier's complete Army record. You always took this with you when transferred from place to place. While stationed at a permanent duty station, the company office or headquarters kept it.

An assembly line of clerks were sitting behind a row of tables, checking orders and records as you progressed through the gauntlet. Were the orders correct in assigning you to the 101st? Were you actually who the papers said you were? They were looking for any problems that might keep you from being assigned to the worst possible assignment ever imagined.

The last soldier in the gauntlet was a sergeant who looked like he was in charge. There was no smiley face; just a frown, piercing eyes, and probably the power to determine your life for the immediate future. He was the one person making the decision about your destination. So, I made a proposal in the form of a question. Jeopardy's Alex Trebek would have been proud of me.

Remember, I am infantry, 11B20 in MOS code, 11-Bravo in soldier jargon. From the draft notice forward, I was going to do everything my government asked, however I recognized quickly that you could make the Army what you want it to be, sometimes.

The sergeant was flipping quickly through my folder, I cannot image he ingested much detail, so I offered; "You don't any place for an 11B with a secret security clearance do you?" My plan was to get his thought process off an infantry assignment and maybe into a military intelligence unit or somewhere that required security clearances. The missile detachment assignment required everyone to have a secret clearance, which took time and entailed costly backgrounds checks. Moreover, not that many soldiers had security clearances. I reasoned this would be a valuable asset.

My offer did not overly impress him, though maybe just taking the initiative altered his thought process. He asked if I could type; do you see a pattern here? I quickly responded with a confident answer of 60 words per minute, those repetitive finger exercises in typing class actually took. He looked at me with doubt and muttered something under his breath.

He called on one of his men to give me a typing test. At least things were moving in the right direction. We went behind the wall of the gauntlet room into the cavernous building.

There was bad news and good news about the typing test. First, it was on an old, well-worn, battle weary manual typewriter. The good news was the text for the test was simple; the word and letter cadence was such that even the poorest typist could score pretty well. The test looked official; it included a handbook and form number. This was probably part of clerk training back in the States. And, to my benefit, the specialist giving me the test did not grade for accuracy, just the number of words typed in a given time limit. The test listed the number of accumulative words down the page in the margin. The results? I typed 100 words per minute! I never left the building. Sgt. B. assigned me to his group. He did have power. Not FUBAR.

~ ~ ~

A new chapter of chaos now opened for me. I soon became a member of a group called the Personnel Management Branch, (PMB) of the 101st Admin Company. As the events that led me to this group revealed, this branch, or I should say the sergeant, made all the decisions of which enlisted men went to which individual units in the 101st Airborne Division. There was a system; he did not act without input, yet he still wielded significant influence over every combat unit in the 101st.

Company commanders from infantry, artillery, and transportation battalions would submit requests for a number of men, with specific MOSs, and rank. During the Vietnam War, combat divisions and brigades were initially sent in whole, but remained in-country indefinitely. This is in contrast to the manner that whole units were rotated in and out of Iraq and Afghanistan. While a division stayed in Vietnam for years, individual soldiers rotated back to the states after one year, or when their enlistment was up.

A constant replacement process was required to keep the division at full strength, or as close as possible. When the PMB group received a request for soldiers, the aforementioned sergeant would simply go through a roster of in-coming replacements and draw a line down the page to designate where they would be assigned. It was my boring job to type new orders designating these assignments. Since this was before computers were used for such detail tasks and before copier machines were developed, the orders were created on mimeograph sheets and then reproduced by the hundreds.

An old adage attributed to Napoleon Bonaparte states that "an army marches on its stomach", meaning the logistics of providing food was critical to the success of an advancing Army. In the modern era, an Army would grind to a halt without the proper paperwork completed, in triplicate.

~ ~ ~

I was not welcomed with open arms in this group. As I pointed out in the last chapter, I was only 184 days away from discharge upon arriving in Fort Lewis for deployment. By this time, I was well under six months from discharge and going home. It did not help that the platoon sergeant wanted to promote me to corporal, giving me rank over others in the squad. This would have been a mere formality, because corporal was an E-4 as was my current Specialist 4th Class pay-grade. However, since I had more 'time-in-grade' than the other Spec 4s, I had seniority.

Some of my fellow squad members were in-country for several months already, and I would be going home before them. Of course, it was not my fault, Army chaos created this FUBAR condition. Nonetheless, the situation was awkward. Soon this would pass however, and we all became friends.

The rest of the squad, about ten men, and indeed other squads were a strange mix of people. Yet, they did reflect the current state of the Army and the draft that put most of us here. I was drafted as a 19 year old with no deferments. Draft deferment could be granted for college attendance, hardship, some farm jobs, and a few other reasons. The Selective Service draft arguably caused college enrollment to spike in the sixties, with many young men attending college strictly for a draft deferment.

A student deferment was only good for four years however; many students believed the war would finish within that time. Much to the dismay of many new college graduates, the war continued, and they were eligible for the draft once more. FUBAR.

The education level in our group included graduates in anthropology, geology, business, and pharmaceuticals. Many of them concocted their own plans for moving out of their present situation.

Nearly everyone in the PMB group was an infantryman. The TOE structure was ignored again. Some of the men had combat experience and were transferred to the 101st when their divisions were dismantled. Many of these soldiers probably joined this group in a similar manner as I did. However, all of us still maintained infantry MOS classifications. Unlikely as it was, everyone could still be assigned to front line infantry units.

Even though we were on a camp consisting primarily of administrative functions, we conducted our own perimeter defense. A rotation schedule put each soldier on the berm line encircling the camp; in bunkers, towers, and firing positions once or twice a week. This was a nighttime task; six men to a bunker who rotated throughout the night with two men awake and on guard at all times. Bunkers were spaced about 50 yards apart, with 40-foot tall towers standing behind every fourth bunker. This was not difficult or necessarily dangerous duty, but it was not something you requested.

In their individual creative ways, several men tried to get out of this setting. The geologist relentlessly tried to get a direct commission appointment to lieutenant just because he was a college graduate. Direct commissions were appointment to officer rank, bypassing most Army training, but these appointments are usually granted to doctors, nurses, lawyers and other advanced educational levels; these officers would now practice their profession in the Army. The logic escaped most of us why the Army would need a geologist. It escaped the Army also.

Another soldier, the anthropology graduate, was anxiously awaiting approval for his conscientious objector (CO) application. The conflict in Vietnam created a much wider awareness of CO options and a larger pool of people tried to gain the classification.

CO classification was usually a function of the Selective Service and was applied to a person's draft classification if granted. Application for conscientious objector classification could encompass a number of reason, most commonly religious or moral grounds against killing another human being.

If a draft-eligible applicant received CO status, this did not relieve the individual of responsibility or service to his country. The conscientious objector draft classification of 1-O specified "Available for civilian work assignment." This meant they could be assigned to a work program by their draft board for a period of 24 months.

The current Selective Service Alternative Service Program would attempt to match a person's skill with local companies conducting business in such areas as conservation, caring for the very young or very old, education, or health care.[6]

It was not clear how an already-serving CO would be handled. Maybe he would be discharged and serve out the remaining time of a 24 month commitment in a civilian job. He was still in the Army as an infantryman when I left.

~~~

The living environment in Camp Campbell was above average compared to combat unit camps and of course life on a firebase. To make it clear, this author respects the conditions combat troops endured. However, we still liked to complain, whether it was about the monsoon rains, no hot or even running water, and of course the food. It is a soldier's duty to complain. Many times we skipped the mess hall evening chow, and heated C-Rations or hiked to an Air Force EM club, which required traipsing over rice paddy dikes, for bar-food and drinks. Our mess hall menu included rabbit and mystery meat too many times.

The Air Force clubs always had the best food and good brands of beer and whiskey. We couldn't help but think it was good to be the first link in the logistics chain. As pallets of food and beer would come off transport planes, we deduced that Air Force personnel siphoned off the best stuff for their use. Budweiser for the Air Force, Carling Black Label for the Army. Coca Cola for the Air Force, Fresca for the Army. FUBAR.

Our own enlisted men's club, of rustic construction, served beer for a dime a can, on Thursdays beer was only a nickel. There was no food, PX, sandwich shops, or fast food vendors in camps this size. We drank our dinner many nights.

Occasionally our section sergeant would trade for steaks and we staged a regular backyard cookout. I am not sure what this career sergeant first class
~~~

might have used in trade, but he was in charge of the branch making personnel assignments. There were shortages of everything, including troops; I am sure more than a few battalion commanders were willing to part with some good whiskey or steaks, which their quartermaster had finagled, for some extra or specific men. It is the way of the Army. FUBAR.

~ ~ ~

Drug problems increased as the war wound down, this has been well documented.[7] In late 1971, the Army even tested every soldier for heroin addiction before rotating them back to the states. 8 Besides boredom and futility spreading through the Army, drugs were easily acquired and cheap. I was naïve about drugs before arriving in Vietnam; they were not prevalent in Germany, or at least not with crowd I hung out with. I smoked my first marijuana joint while on leave before deployment to Vietnam and was not impressed with the experience. Therefore, I was a bit surprised at the common and open use in Vietnam.

Pot at this time sold in the U.S. for around $8 to $12 a bag, a sandwich size plastic bag, but certainly not full. One ounce was the quantity, you could measure it somewhat visually with two-fingers; hold two fingers sideways across the bag and that should be about an ounce.

While marijuana was generally $10 a bag in the U.S, in Vietnam, it was $20 a bag. But different bag! This bag was a medium sized paper grocery bag, and it was nearly full. There was always pot available, though I do stress that it was not abused by most of the troops. Certainly some would smoke every night, but not most soldiers. I was basically a social smoker, motivated by peer group pressure. Never did enjoy it as much as some smokers.

Soon, I stopped smoking pot for reasons that pot smokers, past or present, probably understand. First, there was no place to get munchies. For the uninitiated, smoking marijuana creates a craving for food, especially sweets or chip or tacos; just about anything. On small bases like ours, when you craved ice cream, cookies, chips, and late night pizza, the opportunities to find them was nonexistent. Further frustrating the situation, scrounging through a collection of C-Rations did not uncover any perfect munchy-food. Heated C-Rations of spaghetti, ham and beans, or beef stew were not the preferred snack.

Secondly, Vietnam is not the place to smoke pot if you are the type of smoker that becomes paranoid easily. Every little sound is blown out of proportion. It was particularly bothersome if we decided to visit the Air Force EM club for something to eat late at night, since getting there required sneaking through a break in a fence, and walking on rice paddy dikes at the very edge of camp; close to no-man's land. FUBAR.

Shortly thereafter, for unknown reasons, I quit smoking cigarettes also. The two smoking habits were not connected; the latter was just not enjoyable in the climate.

~~~

Pot was not the only drug easily obtained in Vietnam, though my observation of this episode occurred just once.

The only PX available in our area was back at Camp Eagle, the division headquarters. I wanted to see what camera accessories they might stock since I recently purchased a new camera through a mail order catalog. Our company clerk made the trip every day, shuttling paperwork back and forth; so, I hitched a jeep ride with him to the camp, which was only about five miles away, just off the main highway.

As we turned off the highway to a dirt road leading to Camp Eagle we had to cross some railroad tracks that were built up high enough to not flood during monsoons, and the crossing was rough. When we slowed down to a crawl, a young Vietnamese kid came running out of the reeds toward the jeep. I was a first concerned, I could not see a weapon, but that did not mean he was not concealing a grenade. However, the company clerk seemed quite at ease. The young boy yelled out "Hey GI what you want?" The clerk calmly replied, with an experienced tone "three vials of coke."
~~~

With that brief exchange ended, the boy ran back into the reeds and we drove on. I said nothing; it was none of my business. We continued on to Camp Eagle, the driver dropped me off at the PX on his way to headquarters. Within a few minute of browsing the barren PX, I could see this was a wasted trip, for me.

It was naïve on my part to think they might have cameras and accessories. Camp Eagle was a remote headquarters, with dirt roads and no amenities. It was nothing similar to large bases such as Da Nang, Cam Rahn, Long Bien, or even China Beach, which offered swimming pools, bowling alleys, and well-stocked PXs.

Disappointed, I walked to the division headquarters building to meet up with my ride. He was not long and we soon started the return drive.

As we approached the same railroad tracks, the enterprising young boy came out of the reeds again with three vials of something. These were not the small drug vials typically used for individual coke deals, each capable of holding a gram of cocaine. These were typical four-inch tall test tubes from chemistry class. From what I could see, each one of the test tubes was nearly full with white powder. I had never used coke, or even seen it; I assumed that was what it looked like. The ease that this transaction occurred was surprising to me, but then my real astonishment happened. As the boy handed the vials to my driver, he demanded "three dollars each." Wow, even though I knew nothing of the trade, I was quite sure that was very cheap. FUBAR.

South America produces the majority of cocaine; the Andes mountainous are particularly suited for growing the cocoa plant. Heroin however, which is derived from Asian-grown poppy plants, was easily available throughout Vietnam. It was reported that much of the cocaine sold to unsophisticated users was either heroin or cocaine cut, or mixed, with heroin. Cocaine does not create the near-immediate physical addiction[9] that heroin does; this lead to many unsuspecting U.S. soldiers becoming inadvertent heroin addicts. In May 1971, a congressional visit and report from Vietnam claimed that 15% of U.S. troops in Vietnam were addicted to heroin.[10] Double FUBAR.

Besides the fact that heroin was easy to find and purchase, the question always raised is: why did troops turn to drugs? The detailed answer is different for each soldier, though there are some common threads. These threads are not drug-specific, since soldiers turned to excessive drinking. However, it was the overall conditions, a general lack of discipline, and low morale that caused the breakdown.

Boredom is one of the worst military enemies. As wars slowly come to their ends, fewer and fewer mission-direct activities keep the soldier occupied. Extra time on someone's hands has always been blamed for mischief, bad habits, or worse. This is a common problem at every age, even children. "Idle hands are the devil's workshop" and variations of this timeless phrase go back to biblical times. This is true today as it was two millennia ago.

In conjunction with boredom, questionable orders and silly regulations are frequently issued; all meant to keep idle troops from getting into trouble. Many times this exacerbates the problem by creating resentment towards the system and leaders. For example, as the war was winding down, we were not allowed to fire off flares while standing perimeter guard at night. Even if we thought, we saw someone in the 'wire' in front of the bunker. We were required to call the guard duty officer and ask permission. He may or may not have been at the other end of the radio if he was checking other bunkers. This made for nervous times.

Last is futility. Why are the soldiers there? What purpose are they serving? Many times when the mission is nearly over, there seems to be no sense of accomplishment. The end of World War II was clear; exact dates of surrenders produced VE (Victory in Europe) Day and VJ (Victory in Japan) Day. An enormous sense of pride and accomplishment swarmed over the country and service members. That sense of victory has not been present since the 1940s, though numerous wars have been fought since. FUBAR.

I do not bring this up as part of a discussion on conflicts, but as part of understanding troops under these conditions towards the end of deployment. I have no answers; as these types of situations and hopeful solutions are the responsibility of politicians and officers way above my pay grade.

~ ~ ~

Besides trying to change duty stations, change MOS, or trade for souvenirs, another pastime of soldiers is to make money. The Army pays very poorly. I was married, with most of my pay sent directly to my wife in the U.S. I drew $40 a month in Vietnam, used mostly for beer, soft drinks, and cigarettes until I stopped smoking them. I was already a camera enthusiast, learned how to develop film and photos in an Army service club in Germany. Everyone was buying new advanced cameras in Vietnam and high-wattage stereo systems to send home. My budget did not provide for any of those.

Wherever a demand exists, a solution will rise to fill it. This is capitalism at work.

Our camp was more than a mile across, surrounded by a perimeter of bunkers, multiple rows of barbed and concertina wire, and a series of guard towers. The complete camp held other units than just ours, including a transportation company, an Air Force radar installation, and warehouses. Our company only guarded a portion of the perimeter, which included ten bunkers placed about 100 yards apart, each bunker requiring six men. The perimeter was fully staffed at night, while a smaller number of soldiers patrolled the area during the day.

Each company enlisted man drew guard duty weekly, sometimes more often. Some men really hated this duty. The company clerk posted a weekly duty schedule that called for about 60 men each night. You could count on seeing your name about every five or six days.

Each night, after evening chow, and before dark, the men listed for guard duty met up at headquarters. The nightly guard officer briefed us about any special instructions or warnings, and gave us specific bunker assignments. We also had to sign for a box of hand grenades for each bunker, and of course turn in the same number the next morning, barring any legitimate use. Grenades were tightly controlled since the practice of 'fragging', or lobbing a grenade into the hootch of someone not liked, was all to frequent. An M-60 machine gun was also issued to each bunker.

With the weapons and grenades signed for, along with our own personal gear including an M-16, ammunition, and flak jacket, we walked the half-mile or so to the bunkers.

You never experienced a good night's sleep, as you rotated duty with two men up and four off throughout the night. Sleeping was on hard wood planks inside the bunker and you had to watch out for the rats. The company had one unofficial dog as a mascot, a good-sized mutt name Brutus; you don't name a small dog Brutus. Some men would bring treats for Brutus and try to keep him in their bunker since Brutus liked to chase rats; he would keep them out of your home for the night. The treats were usually something from the mess hall, like the mystery meat that showed up a couple times a week. Brutus was the only living thing in the camp that liked the mystery meat, which the cooks generously called baked steak. The dog was not stupid though; he would take your treats and then trot off to other bunkers to see what they offered.

Then there was rain. One of my favorite scenes from the movie *Forest Gump* included the voice-over line by Forest as his squad was on patrol:

"One day it started raining and didn't stop for four months."

At the beginning of the sentence, the squad was walking in dry weather, by the end of the sentence they were obscured by a curtain of rain.

Welcome to the monsoons. During the monsoon season, everything was wet all day, every day. This included the firing position, modified foxholes of previous terminology, outside the bunker. The nature of these positions was they filled with water when given the opportunity. They were dug into the ground a couple feet, then a short wall of sandbags were constructed in front of the holes. Wood planks lined the bottom, but you still stood in water, and of course, rain was pelting your poncho, which did not keep water from running down your back, or keep you dry from the knees down. Someone was supposed to be in the firing position with the belt-fed M-60 machine gun at all times. You could not leave the position, and if the guard-duty officer came around during the night and found the position vacant, FUBAR.

Needless to say, a lot of men did not like guard duty. Now I did not like it either, but recognized a solution to my money shortfall. I began taking other soldier's guard duty in addition to mine for a fee, I did not invent the service, enterprising soldiers from years past probably came up with idea. Rates would vary between $20 and $40 a night, depending on weather, special activities such as a USO show that you would miss, or a holiday.

The only trouble with this scheme, other than the misery of guard duty, was that you had to work the next day if you took it for someone else, answering for them during check-in. When your own name was on the guard duty list, you had the next day off to sleep or whatever. However, since taking the place of someone else was not official you had to work. Doubling the insult, the guy who did not stand his own duty, got the following day off. FUBAR.

It was my choice to engage in this entrepreneurship. I endured the extra guard duty enough times to get a new top of the line 35mm camera and send home a fantastic stereo system from the PX catalog. I still have the speakers.

There is an old adage from many wars ago that in a foxhole most men will eventually get religion or turn philosopher. I was not confronted with the ultimate need to decide either way, but during one miserable night of rain and exhaustion-induced hallucinations of something in front of the bunker, I promised myself I would never complain about anything again if I made it home. I have pretty much lived my life by that proclamation.

~ ~ ~

The nature of war is that one force is usually fighting in a foreign country. This creates a wide variety of challenges, including the dilemma of whom do you trust?

For military purposes, zones or regions of South Vietnam were designated as Corp areas, numbered I – IV. The northern most I-Corp sector included the cities of Phu Bai and Hue, and reached all the way to the DMZ, which was only about 50 miles north of our camp. This area, which included the 101st, was not secure enough where as U.S. personnel could go into Hue on their own to restaurants, bars, or take in sightseeing of the historical city of Hue. However, occasionally an Army service organization would arrange an escorted, group tour to the Citadel, and other points of interest.

To be clear, these tours were not conducted in nice air conditioned buses; we were packed into the back of an open 2.5 ton truck, with Military Police escorting us in front and behind the convey. Plus, how many vacation tours have you taken with an M-16 slung over your shoulder?

This opportunity was still a nice diversion from everyday boredom. As we traveled the main highway to Hue, we were frequently met by 10 to 12 year old children as they rushed to the side of the road, waving, and yelling greetings to the GIs. I was not sure if this habit developed from receiving candy or other treats thrown from previous convoys; it may simply have been a Pavlovian response, or did they truly welcome our presence?

After passing these cheering fans, we turned around and watched them as we drove away; the cheers turned to jeers, and the nice waves turned into one-finger salutes. You quickly learn that trust is shallow and fleeting. FUBAR.

~~~

I managed to finagle a short R & R trip to Bangkok shortly before I came back to the states. I had been in Vietnam nearly six months, and some good dinners, whiskey, and clean sheets sounded very nice. In addition, I was planning to buy some camera accessories and presents at good prices. Flights
~~~

to Bangkok were free from the military and living expenses were very low, I think one hundred dollars for five days.

Flights with-in Vietnam were always susceptible to changes, as were charter flights out of the country. I traveled to Da Dang with no problem; this is where I would catch a charter jet to Bangkok. For some reason my flight out of Da Nang was delayed overnight, so many of us tried to get some sleep on the concourse floor in the airport. Passengers throughout the airport were nearly all military, including Vietnamese soldiers. I noticed there was a group of them nearby, but that was normal.

I had recently purchased a new camera through the PX mail-order catalog: a 35MM Minolta SLR. This was a top of the line model for the time, and was my first serious camera. The camera included a form-fitting case and leather strap. With my arm placed through the strap, and the camera tucked next to my body, I stretched out on the concrete floor, head on my bag, and tried to catch a few winks. I was not alone; numerous soldiers were around me, all doing the same thing.

After a short period of time and actual nodding off, my arm seemed to fling violently into the air. My initial thought was what the . . . ! Then I realized that my camera was the target, and someone had just yanked it out of my grasp. I jumped up and saw one of the Vietnamese soldiers running away with my camera. Yelling for someone to call the MPs, I chased him down, grabbed his arm and spun him into circles like a dance. After a couple revolutions, he let my camera go from his other hand and it flew about 20 feet. I thought the camera was busted for sure, but still let go of him and went for the camera. I retrieved it, and continued to chase the thief. He rounded a corner; once I turned the corner however, he was nowhere to be seen. Just as well, it was either useless or foolish to chase him.

The melee was somewhat surreal. This event aroused no one else, U.S. or Vietnamese. Everyone went back to his or her business or sleep. Amazingly, the camera was fine, though it had a nice dent in the top that I was proud of; it had its own battle wound, but survived.

The sad thing was the Vietnamese soldier was on our side, or we on his side. The opportunity to snag an expensive new camera overcame any sense of loyalty he might have, and certainly eliminated any trust I might have. FUBAR.

~~~

Both the above FUBAR moments were trivial in the big scheme. Instances of innocent looking kids or women lobbing a grenade at U.S. troops were well documented. And, in the Afghanistan war, Green-on-Green attacks are commonplace; allied Afghanistan soldiers attacking NATO forces.

Whether the events are trivial or deadly serious, trust is frequently a casualty of war. FUBAR.

~~~

Chapter 10 – Short-Timer

All air assets were invading Laos, so some of us hitched a ride through the mountains on the back of an empty flat-bed tractor-trailer to the Da Nang air base for the trip home.

~~~

Nowhere is the term 'short' used as much as in the military. A person can be described as short or a short-timer. It has nothing to do with physical stature. It refers to your ever-shorter amount of time left in the military. You should not, in good taste or conscience start calling yourself short until 30 days from discharge. Otherwise, that would just be annoying.

I was anxiously approaching short-timer status in January of 1971; my ETS (Estimated Time of Separation) was February 19th. I was still a Specialist 4th Class, not a bad rank, but I wanted to go home a sergeant; it carried a little more stature. A promotion would be possible, or so I thought.

Military promotions are regulated so that a unit can only grant a limited number at a time. The TOE for any unit lists how many troops of a specific rank are called for, and if the troop count falls below a prescribed level, a number of new promotions are allocated to the unit. An allocation for E-5 sergeants was received by our company in January 1971; the allocation consisted of 17 sergeant slots. Seventeen men would be promoted form E-4 to E-5. Maybe I could make sergeant yet.

A soldier's promotion up through E-4 was usually at the first sergeant's discretion, though he still needed an authorized allocation. Anything past E-4 required an appearance before a promotion board for examination of the candidate's record and a series of questions; some tactical and relevant to the MOS, and some grilling on military history.

I received permission to submit my application and began studying all
~~~

the infantry terms and procedures I had long forgot such as general orders for guard duty, Geneva Convention terms, the Uniform Code of Military Justice, infantry tactics, and of course the storied history of the 101st Airborne Division. Known as the "Screaming Eagles" from the division patch insignia, the 101st is a well-recognized and decorated fighting force. Major campaigns included D-Day airborne drops behind enemy lines, the Battle of the Bulge at Bastogne, and Hamburger Hill in Vietnam.

With all the pertinent knowledge refreshed, I was ready for the review board. However, the promotion odds were stacked against me. I was going home in less than 30 days. Promoting me to sergeant might be perceived as a waste of an allocation. Compounding that logic, there were over 100 men going after 17 new sergeant allocations. Yet, I confidently and blindly proceeded.

The promotion board was chaired by a major, and included a captain, two career enlisted men; one was a crusty first sergeant, the kind that typically do not appreciate my value unless they know me, and there was no time for that. I put on a clean uniform, showered, shaved, trimmed my mustache, and was ready for the board. The hordes of E-4s lined up outside an office waiting for their names to be called one at a time. Upon hearing your name, you entered an interview room with the review board sitting behind two folding tables. Our previous instructions were to march sharply into the room and stop three feet from the table, directly in front of the major. We were to snap to attention, salute, and recite "Specialist Clark reporting to the review board sir!" We were to hold the salute, arm up, hand at the side of the head until the major acknowledged it and saluted back.

Reporting to the review board as prescribed, I stood in front of the members with my salute in place, waiting for the major to return the salute and begin the review. The next step was not expected.

The major glared at me, stood up, put his hands on the table and leaned forward as far as possible. He stared at me for a couple of seconds, it seemed like an eternity, and then sat down, still not returning my salute.

He then proclaimed:

> *"Soldier, your mustache does not conform to the regulations of the 101st Airborne Division, therefore I must dismiss you from the promotion board."*

The major then returned my salute. I spun around in my best military manner and marched out the room. FUBAR.

The promotion board was obviously overwhelmed with the number of candidates, and searched for ways to thin the herd quickly. That dismissal was rather quick. By all regulations, a mustache is not supposed to extend beyond the corners of the mouth and is not to encroach or cover any part of the coloring of the upper lip. I guess mine was not regulation, but once out the door I had a smile on my face. I was going home in four weeks.

~ ~ ~

EPILOGUE

The events in this book, or similar versions, could have happened in any branch of service, during any modern warfare or even peacetime, and to nearly any rank, men and women alike.

My speculation is that these types of scenarios exist even today. The Iraq and Afghanistan wars have produced numerous FUBAR moments such as pallets of $100 bills unaccounted for[11], inflated prices for fuel, water, and other goods sold to the U.S. military by our own U.S. contractors[12], and learning the hard way about defending against Improvised Explosive Devices (IED), which have cost our country and service people so much.

I have not filled the book with tragic stories, yet they are not to be ignored; they are numerous. If the M*A*S*H television episodes were nothing but death and sorrow, its tenure would not have been as long and its legacy would not be as large. The tragic stories are in front of us every day if you take the time to see them.

In his book *If Not Now, When?*, Colonel Jack Jacobs (Ret.) wrote:

> *Sacrifice is the essence of military service, even in peacetime. Relinquishing the benefits of doing what you want to do instead of doing what you must do is the crux of service to the community.*[13]

His eloquent words describe the nature of life in the military, whether on the battlefield or peacetime assignment, and the fact that brave men and women accept that role with pride.

~~~

I make two challenges. One is for politicians, military leaders, Pentagon planners, and indeed voters to use an immense amount of foresight and common sense before making military decisions, which have personnel,
~~~

economic, and worldwide ramifications for years to come. Put self-gratifying and self-profitable positions aside.

The second challenge is for veterans, families of veterans, and friends of veterans to take a close look at what shapes our men and women, and think about how we can improve the experience and reduce the pain, sorrow, and costs.

I am proud of my service and have never stated otherwise. However, just as individuals can always improve themselves, we can improve the military experience and effectiveness for all.

~ ~ ~

Photographs

These photographs include the good and not-so-good times, FUBAR moments included, taken between February 1969 and February 1971.

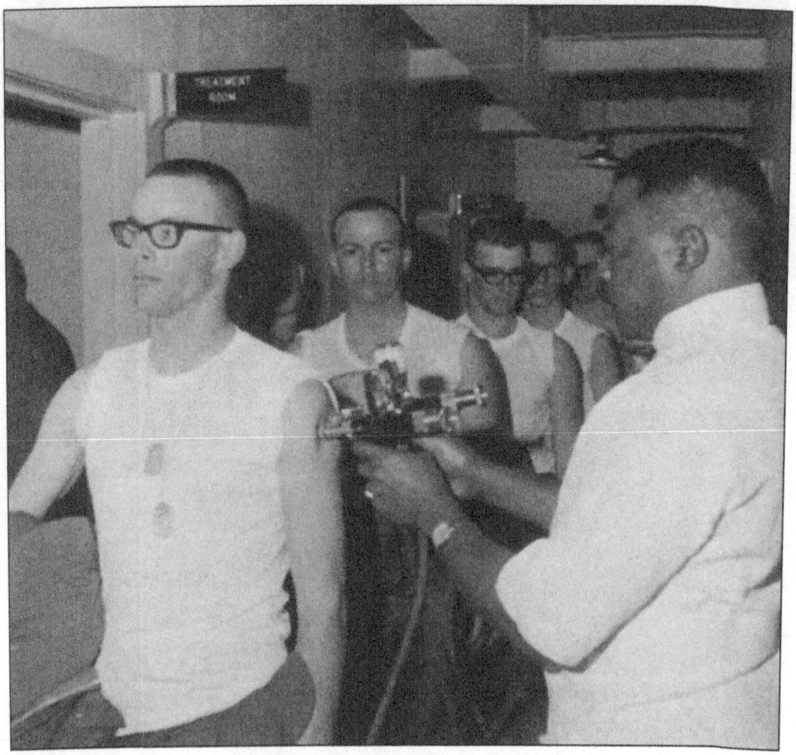

Double barrel inoculations in basic training at Ft. Polk, Louisiana. This photo appeared in the boot camp 'Graduation Book.' Yes, that is the author getting shots from an air gun. The soldier behind me looks more concerned than. That may be because some others just before us had blood streaming down their arms. If the soldier flinched or moved, the air guns blasts would cut the arm, requiring the shots be given again because the serum did not penetrate the skin. [United States Army Training Center Infantry, Fort Polk, Louisiana, 1969]

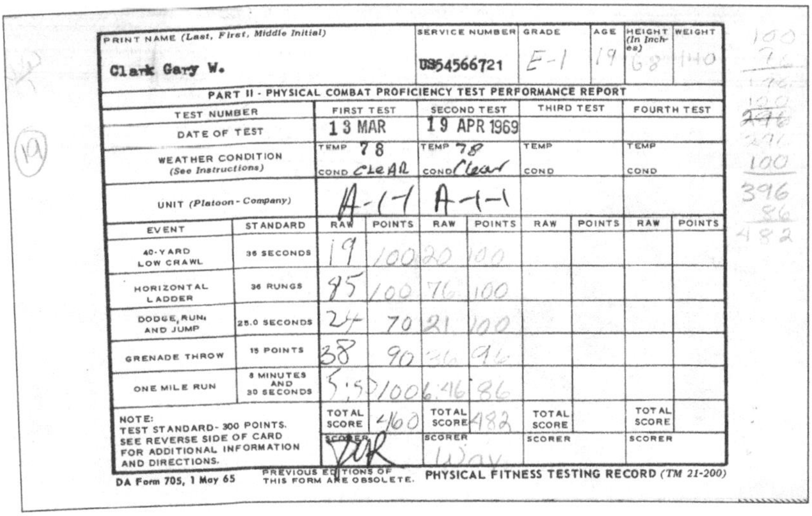

PRINT NAME (Last, First, Middle Initial)	SERVICE NUMBER	GRADE	AGE	HEIGHT (In Inches)	WEIGHT
Clark Gary W.	US54566721	E-1	19	68	140

PART II - PHYSICAL COMBAT PROFICIENCY TEST PERFORMANCE REPORT

TEST NUMBER		FIRST TEST		SECOND TEST		THIRD TEST		FOURTH TEST	
DATE OF TEST		13 MAR		19 APR 1969					
WEATHER CONDITION (See Instructions)		TEMP 78 COND CLEAR		TEMP 78 COND Clear		TEMP COND		TEMP COND	
UNIT (Platoon - Company)		A-1-1		A-1-1					
EVENT	STANDARD	RAW	POINTS	RAW	POINTS	RAW	POINTS	RAW	POINTS
40-YARD LOW CRAWL	36 SECONDS	19	100	20	100				
HORIZONTAL LADDER	36 RUNGS	85	100	76	100				
DODGE, RUN, AND JUMP	25.0 SECONDS	24	70	21	100				
GRENADE THROW	15 POINTS	38	90	36	96				
ONE MILE RUN	8 MINUTES AND 30 SECONDS	5:5	100	6:46	86				
NOTE: TEST STANDARD- 300 POINTS. SEE REVERSE SIDE OF CARD FOR ADDITIONAL INFORMATION AND DIRECTIONS.		TOTAL SCORE	460	TOTAL SCORE	482	TOTAL SCORE		TOTAL SCORE	
		SCORER		SCORER		SCORER		SCORER	

DA Form 705, 1 May 65 PREVIOUS EDITIONS OF THIS FORM ARE OBSOLETE. PHYSICAL FITNESS TESTING RECORD (TM 21-200)

Physical Combat Proficiency Test Performance Report from Basic Training. I scored a 96%, worthy of promotion.

In basic training, everyone is a Rambo. Field training at Fort Polk, Louisiana; I was 140 pounds of lean, mean, fighting machine.

Pershing I Missile outside our barracks on a German Luftwaffe Base.

Practice warhead nose cone and its tracked carrier in background.

Enjoying the sights of Salzburg, Austria from a fortress cafe with my wife. Salzburg, famous as the setting for the movie *The Sound of Music*, was about 130 miles from our base.

Soon-to-be guardians of nuclear warheads – frightening thought.

Darrel B., Harold C., and me at their stand-down camp in the 101st. We were also together in Oklahoma and Germany.

Firing position outside of bunker, main position for the M-60 machine gun. These positions always filled with water during the monsoons.

Motley squad members of the 101st on a hot day in Phu Bai.

Perimeter bunker. The baseball-like backstop in front of the bunker was supposed to detonate RPGs (rocket propelled grenades) before they reached the bunker and make it more difficult to lob a grenade at the bunker. Rows of (barbed wire) and coils of concertina wire were placed in front of the bunker.

Lest We Forget

Three Soldiers statute on the National Mall commemorating the Vietnam War. [LOC Prints and Photographs Catalog, 04696]

Korean War Memorial in West Potomac Park, Washington, D.C. [Library of Congress, Prints and Photographs Online Catalog, 13268]

World War II Memorial on the National Mall in Washington, D.C. [Library of Congress, Prints and Photographs Online Catalog, 04467]

Arlington National Cemetery, Arlington, Virginia. [U.S. Army, http://www.army.mil/article/71288/]

We cannot forget the sacrifices by those who fought in World War I, The Spanish-American War, the Civil War, the War of 1812, the Revolutionary War, and other past conflicts around the world.

Special thoughts and remembrance go out to veterans of the most recent battles in Kuwait, Iraq, and Afghanistan. We will build memorials for them.

Tomb of Unknown Soldier, Washington D.C. [U. S. Navy]

World War II FUBAR

In the 1944 movie Three Privates, Private SNAFU received top billing, though Private Fubar and Private Tarfu had supporting roles.

Rhino-Teeth attached to tank by inventive GIs, a FUBAR situation they overcame. These modifications enabled tanks to punch through solid hedgerows in France after D-Day.

Maps

Vietnam map from July 13, 1971, Wichita (Kansas) Eagle newspaper.

82nd USA Missile Detachment located in Bavarian region of Germany.

SHARE YOUR STORIES

Do you have some interesting, shocking, funny, or not-so funny FUBAR stories? Share them with the world.

I am sure there are plenty of tales, many better than the ones offered in this book.

Share your stories on the FUBAR 2.0 blog, they can be anonymous or you can include your name. The place and time they happened will be helpful, along with the service branch.

Please be tactful, yet truthful, and do not include full names of others in the story. The goal is to illuminate the FUBAR moments that are usually systematic of the environment, and not call out guilty or innocent people.

~~~

For those of you with literary aspirations, FUBAR 3.0 is waiting to written by you. I now pass the baton to the next generation of men and women who gave their service to the United States of America.
~~~

Notes

1. Wikepedia, Rhino Tank, http://en.wikipedia.org/wiki/Rhino_tank

2. Wikipedia, Military Slang, http://en.wikipedia.org/wiki/Military_slang

3. SF Gate, San Francisco Chronicle, August 25, 2007, http://www.sfgate.com/news/article/PBS-conflicted-over-adult-language-in-Ken-Burns-2507231.php

4. Wikipedia, Bell UH-1, http://en.wikipedia.org/wiki/Bell_UH-1_Iroquois

5. U.S. Army Europe, http://www.eur.army.mil/organization/history.htm

6. Selective Service System, http://www.sss.gov/fsconsobj.htm

7. Chang, Daniel, Military Drug Use in the Literature of the Vietnam War, http://dash.harvard.edu/handle/1/8965621, Harvard University, 2008

8. Congressional Report, http://www.npr.org/blogs/health/2012/01/02/144431794/what-zvietnam-taught-us-about-breaking-bad-habit

9. National Institute on Drug Abuse, Cocaine: Abuse and Addiction, National Institute of Health, 2010

10. Congressional Report, http://www.npr.org/blogs/health/2012/01/02/144431794/what-zvietnam-taught-us-about-breaking-bad-habit

11. Washington Post, Iraq Reconstruction Auditor Warns of Waste, Fraud in Afghanistan, http://www.washingtonpost.com/wp-dyn/content/article/2009/02/01/AR2009020102225.html, February 2, 2009

12. The Center for Public Integrity, Waste, fraud and abuse commonplace in Iraq reconstruction effort, http://www.publicintegrity.org/2013/03/14/12312/waste-fraud-and-abuse-commonplace-iraq-reconstruction-effort, March 18, 2013

13. Jacobs (Ret), Colonel Jack, If Not Now, When? New York: Berkley Caliber, 2008, Print

Author Bio

I am a professional photographer, graphics designer, genealogist, and author with a wide range of interests. After discharge from the U.S. Army in February 1971, I wandered around from college to construction work, back to school, and then entered a life-long career centered in the high-tech arena. My jobs varied from computer programmer, to analyst, to trainer, to marketing, to consulting, to author.

I was born and raised in Wichita, Kansas, and have lived in Pine Bluff, Arkansas, Kansas City, Houston, Atlanta, and since 1982 in the San Diego area.

This book has been in my mind for a couple decades and as an outline for more than a year until I finally decided how to present the experiences.

I also write books on vintage photographs, photo restoration, and have a historical drama book in progress. My books can be found on Amazon.com in print and as Kindle eBooks. My Amazon.com author page covers the latest interests I'm involved with and the latest book releases.

~~~

Book Review

If you have the time, you might write a review of this book on the Kindle Books page. Other readers may be interested in your honest opinion of the book. Thank you.
~~~

Made in the USA
Monee, IL
04 November 2026

67384428R00059